.. THE DAY THE BOMB FELL ON AMERICA ..

Winnifred
E. Short
Memorial

J. Morehouse

Also by the author ▪▮ THE JENNIFER PROJECT

THE DAY THE BOMB FELL ON AMERICA

TRUE STORIES OF THE NUCLEAR AGE

CLYDE W. BURLESON

PRENTICE-HALL, INC., ENGLEWOOD CLIFFS, N.J.

■ TO SUZY,
who makes good things
happen. Without her
work and assistance
this book would never
have been written.

The Day the Bomb Fell on America:
True Stories of the Nuclear Age
by Clyde W. Burleson
Copyright © 1978 by Clyde W. Burleson
Printed in the United States of America
Prentice-Hall International, Inc., London
Prentice-Hall of Australia, Pty. Ltd., Sydney
Prentice-Hall of Canada, Ltd., Toronto
Prentice-Hall of India Private Ltd., New Delhi
Prentice-Hall of Japan, Inc., Tokyo
Prentice-Hall of Southeast Asia Pte. Ltd., Singapore
Whitehall Books Limited, Wellington, New Zealand
10 9 8 7 6 5 4 3 2 1

Library of Congress Cataloging in Publication Data

Burleson, Clyde W
  The day the bomb fell on America.

  1. Atomic power. 2. Atomic energy. I. Title.
TK9153.B85   621.48   78-13757
ISBN 0-13-196709-6

FOREWORD

■ The following stories are all factual. The events really took place. All names, locations, dates, and descriptions are as accurate as available documentation can make them. The dialogue is taken from official transcripts and news reports or is a close approximation of what was actually said at the time. Thousands of documents were combined with eyewitness testimony to develop and refine these accounts and ensure their accuracy. Some of the material discussed is still classified by one or another government as Top Secret.

■ Reference is made to nuclear reactors, a study on reactor safety called *WASH-1400,* and the effects of radiation on humans. Where these subjects are cited no additional explanation is required. But for anyone desiring more complete coverage, three brief articles are appended.

INTRODUCTION

■ The day began much like any other. A blazing equatorial sun heated dense clouds of water vapor, stirring them into slow boiling life and causing them to rain down onto the already saturated ground below. Strong winds rolled over the low brown plains, blowing toward the distant blue hills. The land was barren. Only the most primitive plants grew in the highlands. It would have been impossible for man to have lived on the bleak, wind-torn prairie.

■ But there were no men. They and all of their inventions would not arrive for a billion years to come.

■ A Precambrian stream, gorged by the still-falling rain, became a freshet, then a mass of roaring water as it poured down the slope, following channels which were already old landmarks through the rough red soil and rock. As the river moved it gathered energy and collected more earth into its turbulence. Little by little, archeological millimeter by millimeter, the dirt and elements of our planet were moved downstream to lakes and oceans.

■ Part of this primordial perfusion was stopped and held stagnant by small eddies or in pools at the bottom of cataclysmic waterfalls. Year by year, eon by eon, the materials built up. Layered in random precision, they pressed close together. In a millennium or so, the deposits would form sedimentary rock.

■ Then it happened. One atom, wayward and mindless, rocked in its ordained orbit, its nucleus touched by a passing neutron released by another atom not far away. A positively charged proton, agitated by the crude jolt, moved ever so slightly aside, only to be quickly joined by another proton which had been similarly assaulted. Repelled by their positive charges of electricity, they reeled apart and, like a drunk staggering backwards, bumped into still more of their fellows—not to mention a hoard of other neutrons, which, excited by the possibilities of the situation, sped away to look for more of this impact and collision action.

■ A balance so delicate as to be immeasurable was upset. The single wanton act of the wayward neutron caused the entire nucleus of the uranium atom to fly asunder, releasing all of its particles in a hot, infinitesimal flash of energy. Broken and homeless, their stable community destroyed, the remainders of the shattered atom sped outwards, propelled by the death thrust and doomed to move until their force expired or they made contact with others of their kind.

■ The prehistoric rains had washed the ground and leached the rocks, dissolving and carrying along a number of elements and minerals. By chance the water flowed over a site rich in uranium. By still greater accident, layer upon layer of this material, in a relatively pure state, was deposited along the floor of the gorge which had been dug by the water. Finally, it attained a critical mass. The rude neutrons jolted a few protons and caused enough nuclear imbalance to result in the destruction of a number of atoms—a number sufficiently large to ensure, on the basis of chance alone, even more would be affected.

■ The reaction flared. Like a wet match, it gained flaming intensity, then sputtered. An observer on a super-microscopic level would have seen giant suns collide, blaze brightly, then spin off, only to ram the suns of other universes. But the fragile flame caught. A chain of reactions was started.

■ As the impacting and splitting atoms released their energy, heat built up at an alarming rate. In the briefest possible time, the layers of uranium contained a fire of enormous intensity which would burn for hundreds of thousands of years. The molten mass spewed off massive amounts of radioactive materials into our young atmosphere as it charred its way deeper and deeper into the earth.

■ This natural phenomenon occurred at least three times near a place called Oklo in what is today the country of Gabon, situated on the Atlantic coast of central Africa. It is the only discovered happening of a completely natural, spontaneous atomic reaction. It shows us atomic power is not a new or exclusively man-created phenomenon; not solely a product of our technology.

■ Time passed. A billion years or so. It makes little difference, as the earth abides. Then we did what nature had done before us. Only we made it explode. A bomb. Alamagordo, Hiroshima, Nagasaki, Atoms for Peace, then so many more nuclear events the people lost count.

■ We tried to control in a decade what it took the entire history of man to decipher. In a way we failed. In a way we succeeded. But we still know little of the impact of what we wrought.

■ Born in a time of tumult and terror, our newest advance has problems which disturb even the most knowledgeable. But we are trying. And each year we come closer to a full understanding of a force beyond the fanciful dreams of the most imaginative alchemists and sorcerers. Through it all, we too abide. How well, is the question. And where it will lead us is the quandary.

■ The atom offers us an enormous source of energy which our society has a voracious need, because our very comfortable lifestyles are dependent upon the consumption of massive amounts of low-cost power. None of us wants to give up conveniences which have become necessities and give an average citizen a quality of life in many ways better than that enjoyed by royalty only a hundred or so years ago.

■ Our science has defined the nuclear force, contained it, and harnessed it to do our bidding. But there is still some question as to who is really in command. There is a glimpse of unseen fury in every atomic pile.

■ We try to be careful, we try to be safe, and we try to plan for the contingencies which will be the certain result of our eventual mistake or failure.

■ By and large, everyone accepts the fact there will be deaths from man-made radiation. And there will be accidents which release radioactive materials into our environment. The scientists are concerned, the politicians are concerned, civic leaders are concerned, and, because of all the hubbub, the people are concerned. Sometimes. When they are directly affected.

■ An antinuclear group has appeared on the international scene singing songs of annihilation, holocaust, and the end of man. They are countered by the pronuclear army who answer these claims of Armageddon with an "Atomic power is as safe as a powder puff" philosophy.

■ Both extremes are correct but neither is completely right. The tiny atom can provide us with needed energy—and in so doing can blow us all away. But disaster does not have to be a certainty if we manage our affairs properly. And there is the key question: Have we done a good job of our nuclear management?

■ One way to gain sufficient insight to judge is to take a close, inside look at some of the events and happenings which have caused headlines in the world's press.

■ The stories to follow have a common moral, best summed up by an old engineering adage called Murphy's Law:

■ "If it can break or go wrong, it will. At the worst possible time. And in such a fashion as to cause the worst consequences and make repair the most difficult." Murphy was an optimist.

■ Probably the most intriguing thing about these major events of our time is how few people remember them. This alone says something interesting about the society in which we live.

..THE DAY THE BOMB FELL ON AMERICA ..

■ An Air Force flight line is an eerie place in the early morning. In the first gray light most look the same anywhere in the world.

A single line of low structures rose out of an otherwise flat and featureless prairie which extended on endlessly to the congestion of the main part of the base.

The yellow and blue buildings made a natural barrier along one edge of the parking ramp, separated from the rough gray concrete by a narrow asphalt roadway which shone with a tarry blackness. After the service road, there was a scraggly patch of dull green grass. Then, for as far as the eye could see, the concrete continued.

In the early light, colors were hard to discern. And the knife-edged wings of the B-47s, standing one to a pad, each alone, threw strange shadows onto the pitted ground.

Far down the line a truck motor coughed, then roared. A blue silhouette moved along the road, stopping at one of the droop-winged shapes. The sound of the engine lowered and two men jumped out. They moved with practiced rapidity, clicking a nozzle into the fuel filter door, connecting the ground wires to prevent a spark of static electricity from igniting the JP-series fuel the whining pump started to force into the tank. Both men were watchful and they merely nodded to the passing Air Police guard, who marched by slowly, his M-1 carbine slung over his shoulder.

As the light improved, it reflected from the gleaming aluminum surfaces of aircraft. There was a lingering silence broken only by the muffled noise of the idling truck. A lone bird called from somewhere out near the high brown grass bordering the end of the seven-thousand-foot runway.

Things were still but there was tension in the air, a feeling of suppressed power and activity. It was almost as if the noise and

1

hurricane violence of a thousand jet engines had left an imprint of their sound in the rough texture of the concrete.

It was the 11th of March, 1958. The day was Tuesday. The place, Hunter Air Force Base, just a little distance outside Savannah, Georgia.

The first of the spring warmth had brought some color to the southern soil and it promised to be a fine day.

Down the flight line, in the area reserved for the B-47s of the 308th Bomb Wing, the towering empennage of one of the huge aircraft bore the number 876.

A ground crew was working quietly in the chill of the early dawn. The plane had been fueled, its systems checked, and a single nuclear bomb, called a "pig" by the airmen who worked on it, had been loaded under maximum security conditions. The Strategic Air Command was proud of its procedures. Each of its men felt a special responsibility. They knew together they formed the most powerful deterrent force in the history of the world. They were America's first line of defense—and its most feared offense.

A mission was on. But that was nothing new. There was one every week. Training was a way of life in SAC. So much so, some men, fed up with the constant regimen, would ask to transfer out. As with membership in any elite, there were costs. But there were also compensations: promotions came fast and there was a feeling of belonging to the best.

Plane 876 was assigned to fly in a phase of "Operation Snow Flurry," a curious name for a series of field maneuvers which would take the aircraft and its crew, along with others in the squadron, to one of four bases in North Africa. Like all SAC missions, this one would simulate combat conditions. The plane would fly in its optimum trim. The pig on board would be "live," a real atomic bomb, not a simulator. Also according to SAC rules, the pluglike detonator, the single component needed to arm the weapon, would be on board in its locked safety container. Orders were not to insert the device and activate the warhead until a special code signal was given and acknowledged.

By ten o'clock, 876 and her sister ships were pronounced ready. The Form Ones telling the condition of the aircraft had been properly checked and initialed, and the flight line ready-room dispatcher had entered the aircraft numbers in his operational column.

The briefing went well. The officers, wearing the bars and leaves insignia of lieutenants, captains, and majors on the olive green of their flight suits, lolled in chairs as the various elements of the flight plans were presented. Meteorology: no problems over the Atlantic, but some fine winds aloft. Navigation: checkpoints here, here, and here. Radio frequencies for both air-to-air and air-to-ground. Estimated times of departure and arrival. Maps. The assembled pilots and aircraft observers made careful notes, checked watches, and moved the dials of their E6B circular slide rules estimating fuel consumption. Rendezvous points: here and here. Abort instructions. God forbid anyone would have to abort and not fly the mission. SAC was hell on that and somebody's head was certain to roll. But if it became necessary, in the mind of the aircraft commander, then the procedure would be as outlined.

An astounding amount of information was offered and absorbed. The officers, all pilots and air crew, sat through the session smoking cigars and making notes in their three-ring books. An occasional scuffing foot or squeaking chair distracted the listeners but there was no conversation aside from a few whispered remarks when the final destination was revealed. Midway in the room Captain Earl Koehler paid particular attention.

At the end of the questions the briefing closed and the crews adjourned for light lunch. Men who are going to spend the next eight to ten hours in the air at high altitude eat carefully. Intestinal gas is not a joking matter at 35,000 feet, and since there is not much room to stand in a B-47 cockpit, whoever gets it just has to sit on it. It may sound funny, but the pain can be severe.

After lunch a few of the married officers made short phone calls to their wives. They knew their destinations but were unable to reveal them. All they could indicate was roughly how long they might be away. This one looked like an extended trip.

Several modes of transportation were used to move the men to the flight line. A private automobile or two, which would be picked up later by someone with clearance, carried a few. Others took the blue and yellow shuttle bus out to the edge of the air park, then used the flight line trolley to move down to their ready room.

The crews began to form themselves into separate units as they took care of the final details of departure. Maps were folded, lunches and cans of CoCo Malt were packed in briefcases, flight helmets, visors, and oxygen masks were inspected, parachutes drawn, aircraft assignments taken, radio frequencies given a final check, and other minor problems attended to. The men's room did a steady business.

Then, almost as if there were no way to delay the matter further, they straggled out into the bright afternoon sun in groups of two and three, chatting casually. They lugged a small mountain of equipment onto the open car of the trolley, and with a squealing jerk, moved off down the now-noisy and active flight line.

The small tractor pulling the open cars halted momentarily at each parking ramp, and at every stop three green-suited men hurriedly got off, tugging at their gear. Finally, it pulled up in front of 876 and three officers wearing the railroad-track double silver bars of captains took their turn and dismounted.

They stood there a minute in the soft glare of the spring sun as the small tractor engine gunned. With a lunging creak and a cloud of white smoke the vehicle rolled on down the endless line.

The three men moved with great precision. Gathering their leather bags, parachutes, flying helmets, survival packs, and other gear, they made their way across the warm concrete toward the silent plane. One captain carried a locked, marked briefcase containing go-code orders and actual targets.

A ground crewman, assigned to fire guard, stepped away from the dark shadow under the nose of the ship. Above his head, someone had painted a neatly lettered name, *City of Savannah*. He saluted informally and presented the Form One, which the aircraft commander scanned. Folding the gray plastic cover back over the document, he tucked it under his arm.

The three captains moved on to the plane and started their walk-around inspection.

On many of the flight pads the ground check was performed by the aircraft commander alone. But on others, as this one, the crew agreed three sets of eyes were better than one so they worked in unison.

They moved carefully around the towering aluminum structure, checking the condition of rivets in the main spar of the wings, the cleanness of the control surfaces, the dark recess of the nose-wheel bay, and other vital signs which, to their experienced eyes, would tell clear tales about the anticipated reliability of the ship on its long flight.

Satisfied at last, the three loaded their gear, donned the

harnesses locking them to their parachutes, and clambered upward, into the belly of the ship.

The B-47 is not a comfortable aircraft. Its vast size belies its interior room. Because it was built almost like a single-seat fighter, the aircraft commander and the pilot sit under a Plexiglas bubble high up on the nose, above the aircraft observer's station down below in the belly of the fuselage.

There were the usual gruntings and squirmings as they settled themselves, completing the hookups for oxygen and radio contact.

The aircraft commander, Earl Koehler, tried the intercom. From his seat forward, he couldn't see the pilot, Charles Woodruff, who was strapped in directly behind him. "You read me, Charlie?"

"Ah, roger."

"You, Bruce?"

The aircraft observer paused in his task of setting out his maps and calculation devices. "Roger, Captain. Five by five."

The title of aircraft observer was a euphemism for the man whose job it would be to actually drop the bomb. It was revived from the last stages of World War I. In the three-man crew of the B-47 he was charged with navigation, acted as bombardier, and was also the electric countermeasures (ECM) officer operating the gear designed to foil enemy radar and stop retaliation. Bruce Kulka usually had his hands full on a flight. On this one he was really going to be busy because the mission was highly complex. As he worked over the maps and set his radios to the assigned frequencies he could hear the two pilots above him run through the checklist, an item at a time.

Finally the intercom crackled again. "You 'bout got it, Bruce?" The aircraft commander's voice had the flat sound of the Illinois farmland.

"Ah, yeah. I've about got it all set. You ready?" The intercom popped again, and the pilot, from his position in the nose, forward and above the observer he could see the ground crewman holding up the safety pins used to secure the landing gear in the down position. The man held the long steel rods up, one at a time, so the crew on board could count them. The red streamers attached to the ends fluttered lightly in the breeze, making them look like banderillas used in a bullfight. Finally, in dumb show,

5

the sergeant waved the narrow canvas cover normally fitted over the Pitot tube which provided the air-speed indicator information on their velocity.

Woodruff was completing the before-starting checklist and Koehler raised the tower, testing the radio. They replied with the time and the present barometric pressure, to be dialed into the altimeter.

The B-47 is an unusual airplane. Ungraceful on the ground, it flies like a dream. As the first fully operational strategic jet bomber, it was still considered a major plane of the line even though the newer B-52s had more range and outright speed.

Sounds filled the small interior of the ship as the ground crew used the auxiliary starter to begin spinning the huge rotors in the port engines. As soon as they reached critical RPMs, they roared into life. The starboard power plants followed.

From the outside, the droopy-winged aircraft sounded like two trains fighting a tug of war. The noise was muffled in the cockpit, however, and the pilot was conscious only of the thrust of the jets as they gulped huge quantities of fuel.

Jet aircraft are terribly inefficient in the dense atmosphere below 10,000 feet. Their engines, which depend on passing thousands of cubic yards of air through the gaping intakes, burn fuel at a prodigious rate during taxiing and takeoff. So things were planned to get them off the ground as quickly as possible after start-up.

All down the flight line, engines fired and settled into their noisy rotation. Ground crews moved back, driving the portable generators at more than authorized speeds to get away from the stench of burned kerosene which shrouded the field.

Heat waves shimmered from the tail pipes of the now-moving airplanes as they made their cumbersome way out onto the taxi ramps.

Inside *The City of Savannah*, Koehler ran his engines up and kept a close eye on the gauges indicating tailpipe temperature. Behind him, the pilot jiggled the control column to signify he had the ship under his guidance. Brakes off, it taxied forward.

Below, Kulka monitored the various transmissions requesting permission to taxi and lined up his first rendezvous points. He checked the press-to-test lights on the ECM control panel. All the carefully thought out devices, designed to ward off radar detection and send enemy missiles astray, indicated ready, so he reported this fact to the commander, who noted it on his check-off list.

The B-47 was once described as a fuel tank with wings, a small bomb bay, and just enough room to tuck in three midgets to fly it. It doesn't sound comfortable, and basically, it isn't. It takes some getting used to. Every square inch of space around the crewmen is filled with gauges, switches, circuit breakers, plug-ins, and flight controls. There's enough room for a man about five foot, ten, with a sitting height of under thirty-four inches to stretch, slightly, before his head comes into contact with the plastic of the canopy or the thin aluminum sheet covering the side armor plate. The seating position is limited and while there's room to rotate the upper torso, in-flight gymnastics are pretty much out of the question.

Struggling to find satisfactory positions and completing the various checklists, the three men on board became a part of the moving airship.

Cleared for takeoff, the silver and gray bomber moved into position at the end of the main runway. The concrete strip stretched endlessly ahead and even Koehler, in the front seat, could not see the end as anything but a brown line over a mile away. He exchanged comments with the ground control, checked the tower frequency, and on command, started his takeoff roll. The oily smell of the cockpit became more pronounced as he opened the throttle.

Acceleration came slowly as the huge jet engines beat inertia and began to rotate faster and faster. Finally, at about the 5,000-foot marker, the nose of the plane rotated upward and with a shudder the big ship broke clear of the runway. North Africa was more than 6,000 long miles away.

■ Florence, South Carolina, is thought of throughout the countryside as a pretty big place. With a population of about 30,000, it certainly is the largest town between the capital, over in Columbia, and Fayetteville, up in North Carolina.

Florence is known to the people of the area for two things. First, it had a whorehouse. Generations of "good ol' boys" had driven in from Kingstree and Timmonsville and even from as far away as Darlington for an evening on the town. And then driven

7

back, long past midnight, belching bourbon whiskey and filled with tales of their prowess in the squeaky, narrow beds.

Florence's second claim to fame is the railroad. In its large marshaling yards several lines make up their trains. Major repairs are performed when needed on both engines and cars. And judging from the number of people who are employed by the various companies, the need is both frequent and pressing.

Working for the railroad has some rewarding side benefits. Since crews travel across country, a man might work for five days, then have several days off at home with his family. Saturdays and Sundays, the normal times of relaxation for most of America, might be spent on the road, forcing other weekdays to become family times. School schedules get altered for the children to take advantage of Daddy's layovers, and the people in the area make adjustments where needed.

Out from the main part of Florence, a few miles back in the piney woods, the land has a feeling of freshness. It looks, in many ways, about the same as when the English settlers first arrived. Far enough inland from the coastal plain to be a little hilly, the rolling terrain is green with the fresh growth of spring.

Many of the people who work in Florence prefer to live away from the bustle of the noisy rail yards. Commuting is especially easy for the train men. Their hours are as erratic as their workdays, and they move on the roads while most people are asleep or at work.

Mars Bluff isn't so much a town as a place, about ten miles north of Florence. At one time, in years gone by, it was a general store and post office. Little has changed. But the land around the tiny settlement is now owned by people who work in the small city to the south; not too much real farming is done anymore.

Walter Gregg, thirty-seven, married, ex-World War II paratrooper, was one of those landowners. As a conductor on the railroad, his comings and goings were dictated by the needs of the line. Bill, as his friends call him, made good money, and was able to afford the comfortable, roomy, gray shingled house at the end of the short road from the main highway. He lived out where the living is good.

On March 11, 1958, he was at home with his wife, Ethel, working in the garage behind the main house. Walter, aged six, one of a pair of twins, was playing outside the open door. The

8

other twin, Elizabeth; his older daughter, Frances Mabel; and his niece Ella Davis were off in another part of the house, involved in a girlish game.

It was a pleasant day, and not particularly distinguishable from any other. Bill had a few jobs to do around the house, and he and his wife, along with the kids, would cook out later. The conversation would be bound to come back to a discussion of the possibility of adding a swimming pool out back, toward the edge of the property. They had talked about it for some time.

Involved in his task, he didn't hear the sound from the engines of the B-47s. Their fly-over was such a normal event, with the base being located as near as Savannah, it no longer attracted much attention. Especially since number 876 had already attained over 12,000 feet in its on-course climb out. Bill Gregg softly whistled the opening bars of the latest Theresa Brewer hit as he reached for another tool on the workbench.

■ The visored pilot tilted his head slightly to include the altimeter in his scan. Now 13,500 feet. His eyes found the heading indicator. On course. They moved to the rate-of-climb instrument. He touched the trim-tab control on top of the stick with his thumb, and after a second's hesitation, saw the white needle inch up to 500 feet per minute.

All this was done unconsciously. Every motion connected with flying the B-47 had been repeated so many times it was automatic. His mind was filled with early-mission fidgets. He was uncomfortable in his seat. The thought of the bologna sandwich and can of CoCo Malt in the box lunch made him grimace. He tried to relax. He was going to be in the saddle for hours.

Altimeter: 14,000. Air speed: 450 knots. Heading indicator: he corrected slightly; on course.

Suddenly he was jolted from his reverie. The sight of a bright red light on the console hit him like an electric shock. He triggered the radio intercom button on the mike with his left thumb, without removing his hand from the throttle.

"Charlie." There was no special emotion in his voice, but he was tense.

"Got it, Earl. Malfunction light on the electrical bomb-lock circuit."

"Roger that."

9

Neither man appeared to hurry, but both worked quickly.

"Circuit breakers OK."

"Flip it in and out a couple of times."

"Did."

"OK. Bruce?"

"Roger?"

"Got a malfunction light on the electric bomb-lock circuit."

Kulka started wrestling with his shoulder harness. Removing it in the tight confines of the aircraft observer's cockpit was difficult. "I hear you. Want me to take a look?"

"Ah, yeah. And if it looks shaky, pin it."

"Gotcha. I'll be off intercom."

"Right." The aircraft commander's eyes scanned the panel with practiced ease. It now recorded 14,250 feet. Things had taken only a few seconds. The red light continued to glow brightly.

Bruce got free of his straps, and using his arms, scooted himself back toward the narrow crawlway. The dimness of the interior of the B-47 made him pause until his eyes, accustomed to staring out into the sharp light of his console, adjusted. Inching along in the confined space, he moved to the hatch, which opened easily. He squeezed through into the floodlighted bomb bay. In the place where he was standing, the Air Force blue paint had chipped away, revealing yellow undercoating.

With the advent of the nuclear age the bomb load had reduced itself to a single cylindrical unit of polished steel a little over 100 inches long and 30 inches in diameter. It hung, suspended over the fragile bomb bay doors, from a single steel shackle attached to a main spar of the aircraft fuselage. In the bright light it looked like what it was: a gleaming, carefully machined weapon of total war.

Kulka paused for a minute. No matter how many times the ground crew tried to relieve tension by calling it the "pig" or the "hamhock," it always took a second to overcome the feeling of shock and awe in its presence. The bomb seemed to radiate malevolence.

The working space in the bomb bay was tight and he had to move with care so as not to place any of his weight onto the bomb bay doors. Cautiously, he withdrew the stainless-steel safety-locking pin from the gray plastic case on the stanchion above the bomb. The shiny surface glittered in the single bright spotlight, and holding an awkward stance, Kulka fished for the narrow hole with its tip. Once inserted through the shackle, the pin would hold the weight of the weapon even if the electric locks failed.

He was working at arm's length, above the level of the bomb's body, when he first noticed a slight movement of the long steel canister. It was a sort of wobble along the main axis. The pig, which should have been statically secured, was rocking.

Realizing something was desperately wrong, he fished frantically for the insertion hole of the mechanical lock. He changed position, and holding on to a perforated stanchion with one hand, leaned further forward.

There was a metallic "snick" clearly audible over the flight noises of the still-climbing airplane. Then, with a click as the electrical locks failed and the shackle opened, the bomb suddenly fell free.

Horrified, Kulka watched as the weapon struck the thin bomb bay doors. It paused, then its weight and inertia overcame the door locks and they burst open, freeing the weapon into the blindingly bright blue sky below.

It disappeared from the astonished captain's eyes in an instant and before he could react he was fighting for his life. The pressure from air rushing over the skin of the aircraft at speeds close to 500 miles per hour was tremendous. The shattered bomb bay doors allowed the maniacal force entry into the small space where Bruce Kulka was standing and tore at him with unbelievable power, trying to pull him out into space. Instinctively, he'd gripped the stanchion next to his body when the bomb started to fall, and now, standing in the midst of the whistling wind, it was all he could do to hang on. Slowly, an inch at a time, he pulled himself back toward the manual bomb bay door-control switch console where he fought against the hurricane as he wrestled with the safety latch.

In the flight cabin the aircraft commander, who had been watching the red light indicating problems with the electrical bomb lock, jerked slightly. The plane bounced, and a second light, indicating a malfunction with the bomb bay doors, glared ominously.

"What th—" He started to speak into the intercom to the

pilot behind him, then stopped as the plane gave another lurch. This time the feeling was more familiar. "Charlie, did you feel that?"

"Roger. Like a shock wave radiating up from the ground."

There was a long silence and the two men stared at the lights, flying by reflex.

It took most of Kulka's strength to hold on near the place where the manual bomb bay door-activator switch was located. Prayerfully, he cycled the control. The doors moved closed, shutting out the screaming wildness. Carefully, stepping with caution, he crossed the narrow space to the exit hatch and squeezed through into the confined tunnel on the other side. The silence after the mad noise of the howling wind made the normal sounds of flight ring in his ears. He sat for a long moment, getting his breath, then moved slowly back to the crew stations.

"Earl."

The aircraft commander, startled at hearing a voice close behind him, turned in his seat. He wiggled the controls in the age-old pilot's signal. "You've got the aircraft, Charlie." Koehler looked at Kulka.

"The bomb. It broke loose."

There was a long silence. "Oh, my God." Koehler reached for the radio channel selector and dialed in a predetermined three-digit code. His voice was calm as he spoke, "This Garfield 13. I am aborting the mission. Repeat. Garfield 13. Aborting mission."

■ The sound of Bill Gregg's whistling was imitated, lightly, in the air high above the house, as the bomb, free and falling with the streamlined precision intended by its designers, gathered momentum.

On the ground, Gregg first caught a nuance of the sound and his whistle died on his lips as the noise grew louder then instantly intensified into a shrill, piercing scream. Seconds before the note attained its final crescendo, Bill's son shouted to his father, who quickly scanned the sky. Then the end came.

A sudden shock wave, moving quickly ahead of the rumble of the blast, struck with the intensity of a tornado. Timbers flew out and knocked Bill off his feet. The children were tumbled down as if playing some game of their own devising. The house, subjected

12

to immense forces, gave up its structural integrity. Wallboard, plaster, splintered wood, and dust mingled with flying glass to fill the air with lethal shrapnel. The roar of the blast, arriving in its own good time, battered the ears of the fallen people and covered the lesser sounds of destruction.

The shock wave moved onward and out. Cars on a nearby road were thrown off course and spun around. The Mount Mizpah Baptist Church was twisted by the force. Homes as far as a half-mile away were damaged. A new landmark had appeared in the peaceful countryside. A crater, more than 35 feet deep and 75 feet across had been vaporized out of the muddy ground. Tendrils of white steam curled up from the wet black sides. Trees in the surrounding area were stripped of their limbs or knocked flat.

According to Gregg, "It blew out the side and top of the garage just as my boy ran inside with me. The timbers were falling around us. There was a green, foggy haze, then a cloud of black smoke. It lasted about 30 seconds. When it cleared up, I looked at the house. The top was blown in and a side almost blown off."

The screams and sobbing of the frightened, bleeding children were counterpoint to the jumble and confusion of broken furniture, debris, and tumbled interior fixtures thrown out by the blast.

Gregg was shaken, but able to see clearly. "I saw the other children on the other side of the house. About a hundred yards from where the bomb hit."

It was some time before a complete inventory of damage could be taken. At the McLeod Hospital in Florence, all the children but one were found to have only minor physical problems, mostly cuts and scrapes. The niece—as luck would have it, the visiting child—required a total of thirty-one stitches to close a long cut on her forehead. Miraculously, none of the adults were seriously injured. But the house was a total loss.

The bomb had landed and, according to plan, the chemical "trigger" designed to set off the nuclear warhead had exploded with the power of several hundreds pounds of TNT.

But the atomic portion of the device, locked into its safety mode, did not ignite. The components of the weapon were vaporized by the intense heat of the chemical explosion, throwing a wide ring of plutonium contamination around the ground-zero site. This was the only nuclear consequence.

13

Reaction was immediate. As soon as the B-47 landed, its crew was rushed into a night-long debriefing. An experienced Air Force public relations team swung into action to handle press matters on a local, state, and national basis. A special disaster crew was formed and ordered into the stricken town under the personal direction of Major General Charles B. Dougher, commander of the 38th Air Division at Hunter.

The first step was to close the area to curiosity seekers and souvenir hunters. Even though the nuclear portion of the bomb had been destroyed by the heat of the explosion, plutonium had been thrown outward from the crater's center, and presented a potential radioactive threat.

Careful monitoring and a complete physical check of each member of the Gregg family indicated they had suffered no radiation exposure.

A radiological group headed by Major Jack Wilt was dispatched to the crater and it began a "mopping up" operation within hours of the original blast.

Exhibiting true concern for the striken family, an Air Force colonel was sent to the hospital. Years later, the only badly injured member of the group, the young cousin, recalls the visitor rather fondly because he brought her ice cream and talked for a while.

The incident gained immediate attention in the press. The Florence paper reported the situation in a strictly local context, including comments by the mayor. Nationally, headlines blared notice of the event and stories were aided by an official Air Force release which gave scant details but offered assurances there had been absolutely no danger of an atomic explosion. The foreign press went wild.

In England and other NATO countries, where flights by manned, atomic armed B-47s were as common as they were in this country, the comments became very bitter. War and its destruction was still fresh in their memories and they wanted no part of such potential disaster.

The British were especially vocal in their opposition to continued flights over their country by U.S. military planes carrying nuclear armaments. The House of Commons went into full debate and both the Labor Party and the Trades Union Congress came up with demands to halt all such missions.

The Italian, Greek, Dutch, Indian, and other nations, according to their press, remained somewhat cooler than their British neighbors, but obviously no one was enchanted by the idea of the now-proven possibility of accident.

The Soviets, acting quickly to capitalize on the situation, launched a propaganda program using radio as well as print. Their theme, designed to maximize the seriousness of the incident, centered on a single concept. The accidental dropping of nuclear bombs could well lead to the issuance of an attack order, by "some irresponsible official," who would choose to blame the Soviets for the error rather than admit his own mistake. Their theme even offered a cure. "From all this, we can draw only one conclusion," one of their commentators said in a Greek-language broadcast. "An end must be put to flights of aircraft with nuclear weapons. The flights," he went on, "threaten the life of many men and they create a danger for peace all the world over."

The reaction in the United States, however, was pretty much business as usual.

In Florence the mayor, David McLeod, speaking at an open forum attended by Air Force dignitaries, said, "We all realize that we live in periled times, and our nation must be prepared to defend itself at a moment's notice. There are dangers in such defense, and this is one of the dangers."

Neil McElroy, then-Secretary of Defense, also spoke of the perils of our age. "I can only say these are perilous times and that as a part of our security measures, strategic bombers are on twenty-four-hour training. It was one of these bombers that accidently, due to mechanical malfunction, dropped an unarmed nuclear device near the city of Florence." Gregg himself, after being assured the Government would cover his loss, also placed a bright note on the incident.

"I always wanted a swimming pool," he said, "and now I've got the hole for one at no cost. I may open it to the public. Charge them for swimming in uranium-enriched waters."

The Air Force moved as fast as possible with the hearings. They determined the failure had been mechanical in nature, not due to crew error. They immediately ordered all planes equipped with nuclear bombs to manually and positively "lock in" their weapons while on practice combat runs.

Within seventy-two hours the radioactive materials had been scoured from the crater site and all surrounding terrain had been checked and declared free of contamination. A few days later, Col. William Byers had arrived in his role as damage claims officer. To support the payment process, a congressman, John L. McMillan, entered a bill exempting the Gregg family from the federal law which restricted claims to $1,000.

This point brings up an interesting quandary. What if the bill had never been introduced or had failed to pass? Would Gregg's total compensation have been fought on the basis of a maximum $1,000 loss reimbursement? This is a question which still affects us today.

Within the same week it took for Col. Byers to arrive to handle the financial settlements, the press had forgotten the incident and gone on to other matters. It was all over but the actual payments needed for final settlement—which took a full two years to complete. The $54,000 finally received by Gregg was far less than the $300,000 for which he had asked, but with Mrs. Gregg in a state of near collapse the family took the money—and maintained some degree of bitterness. Many years later, the bad feelings remained.

"That bomb came within fifty yards of wiping out my whole family," Gregg is quoted as saying. He also has kept his wife and children away from visiting reporters, sending them inside their new house when photographers come around, which isn't very often anymore.

Among the statements made at the time by apparently responsible people, some are remarkable in retrospect.

South Carolina Congressman John J. Riley of the Second District said the bomb dropping was "something that wouldn't happen again in a million flights." The congressman was probably unaware the military was still searching for a nuclear bomb jettisoned in the sea off Savannah Beach, Georgia, when a B-47 collided with a jet fighter at 35,000 feet only a month before.

■ In the final analysis, the incident boils down to this: mistakes can and do happen.

The best we can hope is most systems will work properly and be operated by individuals with personal integrity who will exercise all possible caution.

The score this time?

One safety precaution worked; the atomic portion of the

16

weapon did not explode. But one "foolproof" device failed and the bomb was released.

After the hubbub quieted down and the Air Force completed its investigation, additional safeguards were developed. Everyone shrugged and things returned to normal. More or less.

2

■ The Mars Bluff incident, if an atomic bomb falling into some-one's backyard can be relegated to the category of a mere inci-dent, is far from the only case of an accidentally dropped or lost nuclear weapon.

Four bombs fell into the ocean off the coast of Spain when a much larger SAC B-52 bomber collided in midair with its refueling ship. The two planes exploded and crashed into the shallow blue waters of the Coastal Plateau. The bombs, knocked loose in the melee, became prime targets for a Soviet recovery effort, so U.S. forces immediately cordoned off the area.

They were not all recovered. Two lay in less than two hundred feet of water and were readily retrieved. A third was knocked from its original resting place into a trench where it fell and lodged on a narrow undersea ledge. Salvage efforts, still classified in nature, were used and the device almost slipped away, nearly missing a drop to even greater, less accessible depths.

The fourth bomb fell into over five hundred feet of water and was never seen again.

On several occasions a "pig" has escaped from a flying aircraft while it was inside the continental United States. At least two cases occurred over the Great American Desert and the bombs exploded on impact. Their chemical detonators blew plutonium over a narrow area, but the atomic warheads remained inert. One officially unconfirmed report maintains a bomber from Kirkland AFB south of Albuquerque, New Mexico, crashed into the Manzano Mountains with a nuke on board. Residents of the area claim the bomb is still there.

The Air Force has a good nuclear safety record, which indi-cates the effectiveness of their basic designs. But, careful as the United States Air Force has been, accidents have still occurred.

19

Although none have had really serious repercussions, there is no such thing as a safe bomb. In theory, any of the wayward missiles could have detonated. An atomic explosion could have occurred. The Air Force has stated the odds at a million to one, but as long as even one chance exists, long shot or not, it can happen.

The Soviets, propagandizing about Mars Bluff, raised an interesting point. Could we be launched into a nuclear exchange by the accidental detonation of a lost weapon? It's a possibility our side has considered at some length. In a highly classified study, this scenario was given serious attention. The conclusion? There might well be, at some time in the life of our republic, a leader with a sufficiently strong sense of insecurity to allow an escalation of hostile events rather than accept responsibility for the accidental destruction of several square miles of populated land. Strong safeguards have been taken against this possibility. The go/no-go system which allows for the arming of our operational nuclear warheads is elaborate. And as fail-safe as man can make it.

The chance of an unintentional or mistakenly started atomic war is almost nil. The word "almost" is troublesome, but as the mayor of Mars Bluff said, we live in periled times.

The Air Force has done all it can to safeguard its nuclear weapons from accidental explosion. Their record is admirable. But they are not alone, as a service, in having their problems.

The Army, for example, has had a number of mishaps with its tactical atomic artillery warheads. These devices, although smaller in size and lesser in power than the strategic bombs, represent a considerable explosive force.

One incident concerned a private first class who, through a series of misunderstandings, was called upon during a European maneuver involving NATO forces to direct from a zone headquarters the movement and deployment of atomic artillery.

By his own admission, the man, whose security clearance was limited to "rumor" status, was responsible for the issuance of orders for the placement and simulated firing of nuclear field artillery shells on a several-hundred-mile front.

The American forces engaged in the operation carried live atomic warheads. When the private was found to have no clearance for this activity, the officer in command dealt with the situation in traditional Army style. He marched the enlisted man down to the proper authorities and ordered, in a loud but somewhat

20

shaken voice, a temporary "Top Secret" authorization. His demand was complied with at once.

The Navy, more deeply involved than the other services in nuclear energy, has special difficulties.

All the military now rely on atomic warheads to an extent far greater than imagined by the public. The Navy, in addition to bombs, artillery warheads, torpedoes, and other explosive devices of various sizes, also has nuclear-powered vessels. According to recent reports, a certain amount of casualness in the operation of these ships and boats has occurred, leading to potentially dangerous situations.

The Norfolk Virginian-*Pilot* newspaper reports several crewmen from the U.S.S. *California,* a nuclear cruiser, cited abuses including the falsification of atomic data to impress inspectors, and overt acts of sabotage or negligence. Examples they offered included an incident in which someone tampered with a saltwater condenser, a situation where cereal was tossed into a steam generator, and a crew game in which one man playfully squirted another with a squeeze bottle full of reactor coolant water while shouting, "You're contaminated." Drug abuse poses another singular dilemma. One entire submarine crew was reassigned because of this problem. The seriousness of these breaches at sea is nothing compared with the potential threat they pose in the tight, crowded confines of a major port.

The armed services of all atomic nations create a very special difficulty because of the large numbers of devices in their stockpiles.

As of July 1, 1976, the United States admits to having 2,124 strategic nuclear delivery systems, including 1,054 land-based intercontinental ballistic missiles (ICBMs), 656 submarine-launched ballistic missiles (SLBMs) on 41 nuclear submarines, and 414 active bombers. We can deliver 8,500 independently targeted nuclear warheads.

Best estimates of the Soviet capability, also as of July 1, 1976, indicate they have 2,404 delivery systems: 1,452 ICBMs, 812 SLBMs on 60 nuclear subs, and 140 strategic bombers. This gives about a 4,000 nuclear warhead capability.

But the 12,000-odd strategic devices are only a small portion of the arsenal. Between the U.S. and Soviet armed forces there are more than 12,000 so-called tactical weapons. Many of these smaller units pack more power than the original bomb which destroyed Hiroshima. All together, among the total number of countries known to have atomic explosive devices for military use, there are probably over 25,000 warheads—and a significant portion of them are on "ready" or alert each day.

The safety record in handling this assorted weaponry has been adequate. The Stockholm International Peace Research Institute, established by the Swedish Parliament in 1966, keeps a running score. According to its 1977 Yearbook, there have been 125 military accidents in the past 30 years. That's one every three months.

Putting it another way, that's a mishap every 90 days for the last 30 years. And in the last 10 years, each "accident" has concerned a device packing the explosive power of more than 20,000 TONS of TNT. (Remember Mars Bluff? A couple of hundred POUNDS made a crater 75 feet across and 35 feet deep. It's still there, twenty years later.) The original Hiroshima bomb only had a 12,000-ton blast. So you can see the newer weaponry is an advance.

Not one of these incidents has caused a disaster. And although some, like the touch tag games played undersea by U.S. and Soviet submarines, could accidentally set off a localized confrontation, the fail-safe systems have proven themselves.

Some of the accident descriptions defy the imagination. A mobile, land-based United States Corporal missile, fully armed with a live nuclear warhead, is reported to have "rolled off a truck into the Tennessee River." A heavy salvage job was needed to retrieve it.

Another strange happening occurred on April 9, 1968, when the U.S. atomic sub *Robert E. Lee* "became snagged in the nets of a French trawler" while on duty in the Irish Sea.

Our submarine and fishnet problems are minor, however, when compared to those of the Soviets, who have had a number of their nuclear undersea vessels snagged by both Norwegian and Japanese fishing boats. And apparently their on-ship safeguards are no better than ours because there are confirmed reports of a fire and the eventual sinking of a guided missile cruiser in the Black Sea in September 1974. Several nuclear warheads went down and there is no record of a successful retrieval operation having been carried out.

There is real disaster potential in the expanding proliferation

22

of atomic devices. Each has a valid use and plays its role in defense. And in today's world we need all the defense we can get.

But it is not until one peruses the full list of nuclear devices which have come into our society that one realizes the extent we have gone down the yellow brick road to a kind of Oz in which atomic power is an everyday thing. It would be difficult, for instance, to count the number of United States citizens given courses during their military training on atomic weapons or power plants. Or the number of nuclear manuals stashed away with service-related mementos.

The proliferation of and familiarity with the tokens of atomic power do not equate to a comprehension of the problems inherent in the force. The military has done a good job in handling its new weapons. But they have contributed to the growing nonchalant acceptance of nuclear energy. And military use places another difficulty before us.

There is no holiday from warfare. Once engaged, convenient stopping points cannot be decreed. The likelihood of combat or precombat pressures placing a strain on the maintenance schedule, say, of an atomic pile, is a very real possibility.

It is interesting to note the only U.S. reactor ever to attain critical mass and explode was one of a military design, intended for a military mission. The SL-1, located in Idaho Falls, Idaho, had been giving trouble for a couple of months. Then, on the evening of January 3, 1961, while being attended by three service-trained maintenance operators, it went off. Reactors cannot blow up like an atomic bomb, no matter how they are handled. But in less than a single unguarded second they can increase their energy output a billionfold, resulting in a possible explosion from trapped steam or a rapid release of highly radioactive material into the atmosphere. SL-1 did this. The three men died.

During times of war, all equipment is called upon to perform at the absolute limits of its reliability. And on many occasions, when the need is great, history shows us almost miraclelike instances when machines were driven past reasonable stopping points. In all probability, a field commander, somewhere, sometime, will have to call on his nuclear staff for more than they believe they are capable of delivering. But this time, if they fail, or if there is a malfunction, or if the weapon won't stand the extra stress, the results will be drastic.

Military planning takes this concept into consideration, and sets specific limits on the use of atomic apparatus. But rules are broken every day by people with less reason than a hard-pressed or besieged general officer who is responsible for the lives under his command.

Likewise, there is no development program sufficiently rapid to deliver new weaponry at a rate fast enough to satisfy a military force engaged in an armed confrontation. Even the requirements by a government for defense preparedness can place time constraints on a project which result in too much being attempted too soon.

It was just such a situation which caused what may possibly be one of the worst man-made disasters in the history of the world.

More than one major nuclear disaster has occurred in the Ural Mountains, inside the borders of the Soviet Union. The term "major" is intended to imply an event of significant magnitude. Hundreds of people killed, thousands injured, and more thousands made homeless by radioactive contamination.

No official comments have been made by the Soviets to any world body, and, as far as is known, no news reports were carried inside the USSR at the time.

First public word of a holocaust came under a somewhat suspect circumstance. A dissident Russian scientist, allied with the community of scientists of the Soviet Union in exile, arrived in London, England.

In his field, Zhores Medvedev is considered to be a first-class genetics man, and it was while he was outlining the growth of the genetics-related fields of study in Russia he mentioned the impetus given to the science and the new importance placed upon it by the government after the accident.

His claims drew immediate attention, and when asked to amplify his statement, the researcher obliged. He not only reiterated the cause of the event but the proportions of the disaster as well.

The possibility there was more to the matter than readily apparent arose when Medvedev was shown to have close ties to an English scientific group with strong opposition to a plan for the establishment of a shallow waste disposal/burial area in the northeastern part of the British Isles.

Shallow burial of nuclear wastes, mostly from power plants and breeder facilities, but with some spillover from plutonium production centers, has been a subject of scientific debate for many years. One group maintains the low-activity wastes can be safely trenched underground at depths of several feet. The opposing camp, with as many references and experiments to the contrary, maintains this is asking for trouble in the long term, and that the only successful disposal method is super-deep bedrock burial, or special container dumping in the depths of the deepest oceans.

Since a material with a hundred thousand-year half-life is at least a part of what is being disposed of in these centers, it's a little hard to come back and correct any mistakes, even though a hundred thousand years is more than enough time to see if one was made in the first place.

Medvedev's alliance with the anti-shallow-burial group cast doubt on his story because, coincidentally or not, it was this same shallow burial he blamed for the Soviet mishap.

The pro-shallow-burial group dismissed his references to the disaster with great disdain. The chairman of the United Kingdom Atomic Energy Authority, Sir John Hill, called the whole thing "rubbish." He maintained the type of low-activity waste being buried "could not possibly give that sort of explosion." He also stated the commission, which is in close touch with its counterpart inside Russia, had never even heard a suggestion of such a happening. He felt, therefore, the report was inaccurate and incorrect.

The story would have died out had it not been for a second scientist, Leo Tumerman of Israel, who suddenly came forward and maintained he had actually driven through a devastated area in the Ural Mountains.

His comments more or less agreed with Medvedev's, but he attributed the damage to a different cause. According to Tumerman, faulty reactor cooling in a manufacturing plant located in an atomic weapons arsenal resulted in an explosion. He did not rule out the possibility the accident might have stemmed from storage of waste matter, but seemed more on the side of the former argument.

Additional controversy came from a report in the Los Angeles *Times* in November 1976, in which two unidentified but separate intelligence sources were given. Both reported the 1957 or early 1958 accident as having been caused by an out-of-control plutonium production reactor in a nuclear weapons complex several hundred miles northeast of the Caspian Sea near the southernmost Ural Mountains.

News of this disastrous event, it seems, had been leaked

across to U.S. intelligence sources for nearly two decades without formal confirmation or denial.

According to the *Times*, our top-secret radiation sensors detected the incident shortly after its occurrence. The knowledge was highly classified and kept secret to avoid showing the Soviets the sensitivity of our atomic detection systems.

The available facts indicate several Soviet nuclear accidents took place between 1946 and 1959. A review of Central Intelligence Agency source material sheds light on the question, and opens the subject up for further expansion.

According to the CIA papers, there were several events. In one, a middle-sized city, complete with a subway, was constructed in a remote part of the Urals for the sole purpose of assessing the damage from an atomic blast. After filling the new town with animals and surrounding it with military equipment, also undergoing evaluation, a low-flying aircraft dropped a 20-megaton atom bomb, scoring a direct hit. This action undoubtedly caused a wide area of devastation.

Another incident cited by the CIA was an explosion at the radioactive materials plant Chelyabinsk-40, located in the restricted area of Kyshtym, which is devoted to atomic research. A reactor at Techa and a radiological institute at Sungul combine with the processing facility to make the Kyshtym zone one of the major nuclear centers of the world. After the blast, stores selling milk and other produce closed, and people existed in a state of near panic from fear of radiation effects.

A third report, unlinked to the other two, indicates there was a severe blowup of an atomic nature which had strength enough to "shake the ground." Trees and other vegetation died following this event.

In the late 1950s, both the Soviet and Western scientists possessed little more than theoretical knowledge about many aspects of atomic energy. Waste disposal was among the less studied areas. In the United States we had our early problems with this difficult phase of the nuclear cycle.

In general there are three types of waste which require three different solutions.

The *low-level* classification makes up the brunt of the material routinely removed from nuclear power stations and plutonium production processes. These by-products do not contain high levels of radiation or significant amounts of long-lived isotopes. Although far from safe to handle, they generally represent no real environmental threat if proper controls are exercised and their disposal is achieved through shallow burial in licensed commercial sites.

The *transuranic* wastes are similar to those in the low-level category but contain or are contaminated by significant amounts of long-lived alpha particle emitters such as plutonium. A great deal more care is required in the handling of this material, and shallow burial is not a proper disposal program.

Finally, the *high-level* wastes, which create the biggest problem, are those by-products created through the reprocessing of nuclear fuel. The history of man on earth could change several times before these materials will stop producing environment-damaging radiation.

In all probability, the disposal of the transuranic and high-level wastes can now be achieved in safety. Long-term analysis of underground multiple barrier storage systems and possible sea floor areas indicates these radioactive products can be kept from our environment.

Thus far, our production of nuclear weapons and our atomic power plants have generated about seven to eight million cubic feet of these radioactive elements. Our weapons program is especially noteworthy because seven million cubic feet have come from this source alone. Estimates vary, but by the year 2000 our weapons industry probably will have thrown off 11 million cubic feet and our nuclear power generating stations about 333,000.

This sounds like a lot, but due to the new multibarrier systems approach—in which the wastes are solidified into a glass-like material, then encapsulated and buried deep in the earth in geologically stable formations such as salt domes or granite bedrock—about fifty acres of land will be needed for this effort. If several sites were selected, the land usage would be under four square miles. It is also possible to place the material into holes drilled into the sea bed at the junction of tectonic plates. Further geological activity will actually carry the substances deeper into the earth.

The low-level wastes, while far less virulent, produce a somewhat knottier disposal problem because of their sheer volume. Now that a clear definition has been made of which materials and radiation levels may be classified as low-level, relatively shallow burial is probably as good a disposal method as any. But it's going to take considerably more real estate than the four square miles needed for the transuranic and high-level wastes.

Much has been done in this area in the last twenty years, and there is no question our current knowledge of do's and don't's far exceeds our understanding of the problem as it was viewed in the late 1950s.

Incidents in which high-level wastes mixed with low-level materials have been buried in too shallow disposal sites have occurred, and in this country, have been detected by increases in heat in the waste piles.

Deep-sea dumping has also been tried by both the U.S. and Russia. Soviet nuclear submarines are known to discharge their low-level wastes directly into the ocean. And programmed U.S. disposal has resulted in some leakage of the low-level variety of material from sealed drums dropped offshore from the 1950s through 1976.

The low-level wastes pose a special problem because of their volume. By the year 2000 the General Accounting Office (GAO) estimates the nuclear industry will have produced over a billion cubic feet of this material. Some of it has leaked from its storage and has been found as a contaminant in streams and water tables. Work is moving ahead on this problem, but the dilemma is far from solved.

The effects of poor disposal technique can be graphically seen from the next story. The damage could have come as Professor Tumerman believes, from a reactor fire caused by improper cooling in the older graphite core units. But faulty waste storage was probably the cause.

In any case, the actual fact of the disaster is not in question. However, since the extent of the death and injury list is open to some debate, for purposes of this narrative it seemed wiser to side with the conservatives. The following account has made use of the eyewitness statements of those few individuals who have visited the devastated area, certain CIA documents, and a limited number of sources which were able to provide, a piece at a time, some insight into the mishap. A small amount of supposition is necessary, due to the confidential state of Soviet nuclear technology. The Russians have been very open with the United States and other countries, on a scientific level, about their work on fusion and advanced nuclear physics. The international community of scientists tends to talk a lot more than many security people, on both sides of the curtain, deem wise or advisable. But as far as the early days of Soviet atomic energy are concerned, there is still a great gray area over practical matters of everyday operations.

■ The upper end of the Caspian Sea lies about as far north as central Maine. Completely landlocked by the body of Mother Russia and the mountainous coast of Iran, it is fed by a number of rivers including the Volga and the Ural, which springs from its high source a few hundred miles away in the mountains that give it its name.

At its northernmost point, in a great flatland called the Caspian Depression, at precisely the point where the mighty Ural flows into the inland sea, is the settlement of Gur'yev. In addition to being a railhead, the town is the southern terminus of the great highway which runs roughly north and south to the city of Orsk and thence northwestward across the mountains.

The land around Gur'yev is actually below sea level and is covered with a wandering marsh stretching miles to the south. The entire area falls into the state of Kazakh, the second largest in the Soviet Union. It is also one of the very richest, blessed with fine rolling farmlands and a more more moderate climate than that possessed by Moscow, to the northeast.

Northwest of the Caspian, toward the Ural steppes, the land is open and bare. No great number of people populate the area and, apart from the main road, only a few all-weather tracks are available to carry modern motor vehicles.

One of the settlements in this cold wasteland is the town of Blagoveshchensk. "Installation" might be a better word. The original village had existed for years as a gathering point for the farmers and traveling merchants. It, and perhaps eight or ten other towns spaced a little over a day's journey by caravan, combined to make up a sufficiently active and populous area to lure peddlers and vendors.

Blagoveshchensk is located on the European side of the Urals. The people of the area, although differing in the ways and customs of life from their French or German cousins further west, were Christians. Their features, while possessed of a somewhat Slavic cast, were clearly differentiated from the more Oriental look common on the eastern side of the great mountain chain.

Life in the little community had continued without change for hundreds of years. The traditions of the people remained constant. Boys courted girls in the same ritual, were married, had children, grew old, and died. The next generation followed their example. Wars and Czars came and went. The village was too small to be fought over, too insignificant to be included in anything other than a cursory political list, and too unimportant to be considered for any projects of consequence.

Until one day in the early 1950s.

A planning group, convened in Moscow to consider the best disposal techniques for the low-level wastes from the plutonium production reactor in the nuclear weapons complex off the main highway northeast of the Caspian Sea, began to consider alternatives.

Trouble had plagued this site since its inception several years before. One reactor, with its pile constructed of graphite, had generated and contained enormous amounts of heat. During one period of operation a critical point was reached, then exceeded, resulting in a skyward-booming rush of incandescent gases. This radioactive flare caused the deaths of several people inside the installation compound and released lethal amounts of radiation high into the atmosphere, where it was tracked by U.S. air sampling planes patrolling the Soviet borders. People in a nearby settlement were exposed to damaging amounts of alpha and gamma particles. Illness and death followed.

The reactor itself was so radioactively hot it was useless. Contamination levels were so high no humans could linger in the area in safety. Work to rebuild the site was out of the question.

Annoyed at the loss of the unit, which had to be covered over by such a huge amount of dirt it resembled one of the Ural foothills along the distant horizon, the Soviet scientists learned what they could from their mistake and plunged ahead into the construction of a new and improved plutonium producer at an adjacent site.

The loss of the costly equipment taught them the value of decentralization and the need to disperse their nuclear risks. An

31

accord had been reached before the special emergency meeting and the decision to use shallow burial techniques to dispose of the low-level wastes was agreed to by the main committee. The question before the group was where the site should be.

The location had to pass two tests.

First, it should be in geographical proximity to the reactor. Not too close, but not too far either, allowing for easy transport of the radioactive materials.

And second, it had to be in a sparsely settled area, near some existing town where labor could be found and the disposal experts housed while quarters were constructed on the burial site.

Detailed maps were studied and Blagoveshchensk was selected. How the specific determination was made is not known to anyone in the West, but the location fit both requirements.

Official inquiries were made through the government, terrain maps and photos acquired, and an engineering survey conducted. The site seemed perfect. Designs were produced for the complete installation, and the order given to start work.

The first the people of the village knew of the plans for the installation was the arrival of a survey crew. They stayed in the area for a few days, working about 12 kilometers from the edge of town in a barren area of public land. Then they were gone. Life went on as usual. The brief interruption caused by the outsiders' presence disturbed the elders, but resignation to the fates and the continued absence of any further sign of activity lulled them into a complacent state.

Winter passed slowly. The dark nights and cold, brilliantly white snowy days blended into one another. The events of the previous summer were discussed, then forgotten in the harsh realities of the long, below-freezing months.

Finally, spring came, and with it the first of what was to become a large construction contingent.

They were polite people and when visiting the village maintained themselves slightly aloof, but always with courtesy. Only one or two incidents occurred, and the girls involved were known to be of a less than spotless reputation anyway.

Explanations of their work to the villagers was futile. While everyone had heard of the atomic bomb, more or less, the nature of nuclear waste and the necessity of disposal was foreign to the country people. Glowing reports in *Pravda* of growing Soviet scientific technological understanding of the atom made almost no mention of such by-products.

Construction on the site took the entire spring and summer. Then, in the fall, all was quiet again. The outside workmen went home and the local men who had been lucky enough to find employment at the construction site were ready for a rest. By the first snows the old people were once more lulled into the previous winter's feeling of forgetfulness. The foreigners had come, and now, as always in the past, were gone again.

But new priority had been given to the disposal effort. Through the freezing cold of winter, work proceeded on the necessary roadways to link the production site to the burial grounds. So by spring, the first trucks started to roll, carrying the accumulated radioactive materials to their resting place.

The end of the winter brought renewed activity at the disposal site. The crews not only returned, they came in greater numbers. The surprise of the elders was evident. Meetings were held and much discussion took place concerning the best possible action. But there was nothing to be done.

The work party this time also included a number of semimilitary armed guards. These men, along with a small technical and scientific group, would form the permanent staff at the dump. The duty would be hard. But between shipments, there would be ample time to work on theory, or attempt to make friends with the female half of the townspeople.

The arrival of the first truckload of atomic waste was unheralded except for official notice to the committee which had initiated the affair.

The people of Blagoveshchensk neither knew of its coming nor cared. The compound was far enough away, as long as the strangers kept to themselves, to ensure the villagers' traditional tranquility.

After the first disposal activity, in which a large trench about ten meters deep was scratched into the earth by a special machine, filled with hot debris, then covered over by a bladed bulldozer, ground surface readings were taken. No indication of leaking radiation was found, so the burial depth was deemed to be sufficient.

Soviet engineering tends to depend on the empirical result of actual tests as opposed to theory. Once the technique seemed to

33

prove itself the trucks really began to roll in. Hot waste which had been temporarily stored was sorted out and allotted for immediate disposal. Eagerness to use the new facility caused carelessness. Not all the matter was of the specified low grade. Since the plant had been jammed with waste containers for some months prior to the opening of the burial site, a certain amount of transuranic and really hot matter got mixed in with the original low-level materials.

The expanding nuclear weapons operation was producing a great deal of throw-off and the demand for immediate removal and disposal was a constant one.

Winter or summer, through snow or clear weather, the trucks continued to roll, using the old main highway and the network of newly constructed roads to service the run between the pile and the dump.

Monitoring at the burial site must have produced strong indications that some of the arriving waste was too highly radioactive to be casually buried along with the lower-risk materials. Official correspondence undoubtedly went back and forth between the site commander and his immediate supervisor. But to no avail.

Everyone was acting under orders and a time schedule which would permit no significant delays. Stalin himself had set a date for the first Soviet test of a nuclear explosive device. That time table had been met, only to be replaced by another outlining the production of even more powerful nuclear weapons.

The plutonium facilities ran on a 24-hour basis producing the strategic material day and night, and the waste disposal problems increased in direct proportion to manufacturing efforts.

Trucks came and went. The deep snows of winter covered the older burial mounds but the new trenches were visible as precise black grooves in the whiteness of the blanketing ice.

Monitoring continued on a regular basis with careful notes taken of the condition of every burial site, from the original trench to the newest in the geometrically placed series.

Less care was exercised in the selection of the various levels of radioactive wastes being included for burial. Even the trained scientific staff made poor decisions in this area. As in the United States, real understanding of the potential danger of the debris from plutonium production was lacking. Everyone knew it was dangerous, but the material seemed so innocuous.

34

More trucks arrived. The original trenching site expanded, in accordance with the basic plan for the installation, and everything looked to be going well. A couple of minor incidents developed among the unskilled workers due to their lack of knowledge about nuclear contamination, but they came to little: minor burns and the need for long, hard scrubs in the showers augmented by light medication.

In town the people relaxed. While there was some interchange between the villagers and the newcomers, great care was exercised on both sides and things had settled down.

A year passed. Then two.

The original burial site, where the initial consignment of hot material had been laid in and covered over to the best of the available knowledge and skill at the time, began, far under the earth, to generate critical amounts of heat. The measurement sensors noted a temperature buildup, but the coldness of the winter days, together with a growing nonchalance on the part of the technical staff, combined to dull their minds to the seriousness of the situation.

The end came with frightful suddenness.

On a bright afternoon, with a minimum of noise, the original trenching site simply shook itself awake and, with a single release, spewed radioactive gas and earth up into the sky in a kind of nuclear belch.

What had happened was as simple as it was devastating.

The high-level and transuranic wastes which had been carelessly mixed into the low-level matter destined for burial disposal had reached a point of critical mass.

The ability of the various materials to interact was known to be so great that during production, enriched uranium might be placed in trays side by side, but never stacked one on top of the other.

The result of the unintentional and ignorant mixing of hot materials was a chain reaction resulting in enough heat to actually melt and vaporize the frozen earth. The ditch had been close to this crucial point for days—perhaps even months. Then the single, final nucleus was shaken apart by the invisible bombardment, and the mass became instantly unstable.

There was no explosion in the traditional sense of the word. The ground shook, but the loud blast was missing.

With more a rumble than a bang, it erupted "like a kind of a volcano," and spewed forth the virulent radioactive gas and debris into the air.

As was normal, a light, lonesome wind was blowing from the still plains toward the frozen heights of the Urals. It was no stronger this day than it had been on thousands of other days since the beginning of man's ability to live in the region. But this time, instead of bringing dust or droplets of rain in towering thunder clouds, it picked up the particles of the still-spewing critical mixture of nuclear waste materials and spread them along a narrow corridor.

The hot particles lifted from the earth and, once free, went roaming with the breeze, away from the hills toward the bare, open countryside and the village of Blagoveshchensk, 12 kilometers away.

An 8-kilometer breeze is enough to notice, but not, in the open country, of a sufficient force to comment upon.

It took over an hour for the first of the deadly material to reach the edge of the town.

At the dump installation, panic abounded. All scintillation counters indicated such high readings there was no way to approach the still-steaming burial trench. And almost everyone on the staff had received doses of sufficient strength to turn the photographic film in their personal radiation detectors black. Telephone calls rattled at emergency speed up and down the wires to both the plutonium reactor plant and the seat of science in Moscow. Indecision due to surprise and disbelief was rampant.

The breeze continued to blow over the area and pick up the emittants. Not as virulent as the first releases, they were nonetheless damaging to humans and they continued to be taken up at a steady rate.

In the village, there was no sign of anything amiss. The people moved around outside their houses going about their business as usual. The air looked no different, felt no different, and smelled no different.

But each breath they took contaminated their systems.

The breeze passed on, and as the radioactive materials moved along, the natural currents of wind dispersed the dust and vapor over a larger and larger swath of land.

One by one the other villages in the surrounding area were engulfed by the deadly atmosphere.

36

As the wind moved the particulants, more and more of them began to settle, resting first on a green blade of grass jutting out from the white snow, then falling to the cold icy earth.

Instructions finally came to the dump site. The open trench must be covered over by a large mound of dirt taken from the emergency pile which made a huge mountain near the center of the installation.

Brave men, some of whom realized they had already faced a lethal dosage of radiation, sprang into action and trucks, backed by bulldozers with huge ten-foot blades, started to work. In a matter of hours the open trench was capped and the release of the pollutants into the atmosphere was stopped.

The seriousness of what had happened was not immediately evident to the scientific community. No word of the remarkable event was passed outside the bounds of a narrow group. Geneticists and other scientists of different technological backgrounds received no information on the disaster. Debates over possible courses of action went on long into the night. The outcome was an order to send field investigating teams into the countryside to determine the extent of the radiation fallout.

The organization of these survey crews took days. No previous thought had been given to the need for such units outside the wire-fenced compounds where radioactive materials were regularly handled and stored. No one had anticipated a disaster requiring the study of several hundred square miles of land would ever be necessary.

Before the first crews were in the field, the early symptoms of radiation sickness had struck the less physically fit. The preliminary signs—nausea, a vague malaise, and headaches—would be followed by the more ominous physical indicators in a matter of days.

The dump site itself was closed. Studies of the buildings and equipment taken by the technical staff assigned to the burial center indicated the area was far too hot for human habitation. Vans outfitted to carry the exposed men without danger of contamination to the drivers were arranged for and the station was vacated.

The plutonium reactor operation, faced with no disposal area, was forced to curtail its output and more time was expended on devising interim ways of handling the low-level wastes to meet previously agreed upon production goals. Failure in this area, which was watched by the highest governmental sources as well as the military, could result in the early termination or ruination of a promising career.

Informed members of the Soviet hierarchy took an expedient view of the matter. What was done was done. The thing was to salvage the reputation of Soviet science. No word of the disaster must leak out. And until the field survey was complete and a real understanding of the seriousness of the incident had been obtained, no one should take any action at all.

Life went on as usual in the villages. The nuclear dust and debris had long since settled from the air. In the places where eddies of wind had performed their random dance, the concentrations were high enough to cause serious injury to animals and humans.

The survey teams were astonished at the levels of radiation they found. Exposed ground, water in wells, snowdrifts on the back sides of hills and hummocks, building walls, and the people themselves—not known for their propensity to bathe during the long, dark winter months—were contaminated sufficiently to cause occasional off-scale readings.

The devastation was complete.

High scientific and governmental circles were now faced with a vexing problem. Medical aid could be sent in for only short periods of time. And the people in the drift area could not be moved en masse without causing the situation to become very noticeable.

Sickness had already set in, and to prevent individuals leaving or running away before a decision was made, checkpoints on the only roads through and around the perimeter were established. The guards were instructed to be humane but firm and to contain the residents in the area. All outside traffic was halted. Even cars intending only to pass through were re-routed on the pretext of a scientific experiment supposedly in progress.

Experts in radiation therapy were called upon to give their best opinions about the potential injury suffered by individuals in the disaster zone. Many of these men, asked for their knowledge as if the problem were only a theoretical one, did not realize they were ruling in a life-and-death matter.

Coldly objective again, the government decided to set up health centers for the least exposed individuals in the areas inside the zone where the radiation was at its lowest. These field hospitals could house and treat the people who had received non-lethal doses.

For those less fortunate, two solutions were accepted. The first, which covered the village of Blagoveshchensk and others in the area of highest contamination, was to cut off all communication to the outside world and allow events to take their course. Massive relocation was out of the question, due to the need for secrecy. And obviously, a hush-up policy was required or the noble scientists of the Soviet Union's top nuclear agency might be found to have been less than brilliant. Besides, the damage had been done. Measures were being taken to examine and more closely monitor the other shallow burial centers. There was no reason to needlessly alarm the populace concerning an event which would never occur again.

In short, governmental policy dictated the people exposed to the heaviest dosages of radiation be left to live or die. Some medical aid would be offered, but no one could be moved. It would be far better for the event to appear as if it had been caused by some unknown and powerful epidemic than for the truth to leak out. Besides, observation would provide valuable information on the effects of high-dosage radiation exposure among humans.

Weeks passed. Daily monitoring of the area, done surreptitiously, indicated no decline in the number of hot spots or their relative strengths. The entire countryside was deadly if a person remained inside its perimeter for a sufficient length of time.

The members of the medical staffs who were treating the less severely injured were rotated and their number was cut to an absolute minimum.

As the patients at the small treatment centers recovered they were transported away from their traditional homes and relocated in nearby large cities. Some were given jobs. Others, still too ill to work, were placed on the state rolls. They had all been told the same story. They were the fortunate ones. What had caused the illness was unknown, but they were the sole survivors of a pestilence which made the Black Death of the Middle Ages look like the common cold. The disease was so strong, the buildings of the small towns in the infected zones had been burned to the very ground in an effort to stop its spread.

This, in fact, was being done. Recognizing the contamination would linger past the memory of any living man, the decision was reached to raze the area. As soon as a family died or was relocated their dwellings and out-buildings went up in smoke. The entire countryside was methodically being put to the torch to create an area of visible devastation which matched the invisible menace to be found everywhere.

During this same period, teams of cleanup men were sent into the contamination zone to bury the remains of the dead and dying domestic animals. Even those which had remained relatively healthy were dispatched and piled into shallow graves.

In two months the work was complete. All of the checkpoints had been turned into permanent establishments, the last of the villagers was gone, and the wind, now free of radioactive particles, was able to blow across an area of about 600 square miles, unimpeded except for an occasional still-standing stone chimney. Even compared with the great deserts further to the southeast, this once-fertile land was bleak and barren. And, unlike the cold, sandy wastes, deadly poison to all living things.

The size and audacity of the cleanup and relocation program was unprecedented, but, as is likely to happen when such a huge undertaking is attempted, it was impossible to prevent people from talking.

The medical staff knew the real truth. They quickly realized they were not treating some kind of contagious bacterial or viral disease. The treatments they were instructed to prescribe were for radiation sickness. Like all humans, they told what they knew to colleagues, friends, and family.

Before the operation had ended, the word was out, making the rounds of the rumor mills. A major atomic accident of some mysterious nature had occurred in the far-off area north of the Caspian, in the fringes of the Urals.

The controlling governmental body could do little to squelch the persistence of the story. The general level of ignorance, due to the newness of the whole atomic scene, provided some protection because most of the population did not have any idea of the possibility of a nuclear disaster on such a scale. Everyone had heard of the bomb, but no one knew much about contamination from other sources.

Some effort was made to handle the matter without taking

official notice of the circulating story. *Pravda* ran an article which, without citing the cause of the prevalent tale, outlined the safety and efficiency record of the Soviet nuclear industry—and touched on the techniques of disposing of low-level wastes and their very fine success record in this area.

No information is available in the West as to whether or not this propaganda proved to be effective. At any rate, no notice was taken of the actual event in the Soviet press or electronic news media.

And within sixty days of the incident, work was back to normal at the plutonium production center.

Normality will be a long time in returning, however, to the wasted land. In June of 1961, Leo Tumerman, now Professor Emeritus at Weizmann Institute in Israel, and several other scientists were driving along the main north-south highway in the foothills of the Ural mountains on their way to visit the construction site of the first major Soviet atomic plant at Beloyarsk in the northern part of the mountain chain.

The car in which he was riding reached a point on the highway where large billboard signs had been placed, cautioning drivers to roll up their windows and proceed as rapidly as possible through the area. No stopping was allowed.

Curious, Tumerman questioned his driver about the cause of the warnings and was informed there had been "a tremendous explosion several years before and ever since then it had been like this."

According to Tumerman, they drove for miles through utter devastation. "On either side of the road there was nothing. An empty land. There were trees and grass, but where there once were villages and herds and industry, there was nothing. Only chimneys remained." The professor said they drove for twenty miles and each side of the road was a vast wasteland.

"It must have been sixty square miles that we could see in the valley, and we were not really close to where the explosion occurred. The entire area was hot, very radioactive. The other scientists were discussing which was more dangerous to eat in the area—fish or crabs. I asked the driver to stop so that I could drink some water and he told me it was forbidden. Only chimneys remained of the towns that were once there."

When questioned about the cause of the disaster, Tumerman

said the accident, as far as he could ascertain, was due to a mishap at a military plutonium deposit "either from burying atomic waste or from processing plutonium to make bombs."

He was certain of one thing: "The explosion was the result of the negligence of officials. They were careless and a catastrophe occurred."

The zone of devastation described by Tumerman is greater than the one outlined by Zhores Medvedev but is in the same general location. A lot of nuclear testing has gone on in the Urals, so it is highly probable there is more than one wasted and now-forbidden area.

Robert G. Kaiser, a Washington *Post* reporter and author of the book *Russia: The People and the Power,* discusses the Soviet nuclear weapons program. According to Kaiser, "Early experiments resulted in tragic accidents, some of them contaminating large areas in the Urals east of Moscow and causing hundreds of deaths, but nothing slowed the Soviet push for atomic weapons." His comments agree with the CIA reports.

There is no doubt about the information obtained from air samples along the Soviet border in late 1958. A major nonweapon explosion had taken place. And there is little question concerning the veracity of Medvedev's story. As a geneticist, he was personally acquainted with several fellow scientists who were stationed inside a devastated area or who did research on the effects of the radiation on plant and animal life.

The facts are simple and damning. Inside the Soviet Union, successful development of an atomic bomb was given top priority, and because its makers acted in both haste and ignorance of the potential problems stemming from the production of such a weapon, a number of separate incidents occurred.

It should be made clear, however, these incidents were not of the type which might happen in the far tamer nuclear power generating station. The plutonium manufacturing process was, at the time, in its early stages, and much has been learned since. In addition, the type of installation utilized for electrical power production is different from a graphite-cored unit designed to produce radioactive materials.

Recent speculation has come from the antinuclear forces to the effect the U.S. government might well have publicly released information about having detected the blowup of the

Soviet plutonium reactor in an effort to cover up the eruption of the shallowly buried wastes, since waste disposal is today a very controversial subject. No verification can be found for this assertion. The articles which appeared in the Los Angeles *Times* revealing U.S. official knowledge of the reactor station explosion are also said by some antinuclear forces to have been written in a slanted fashion to draw attention away from an event which closely parallels today's disposal techniques here in America.

Neither the pros nor the cons waste any opportunity to state their case, and we are bombarded by, on the one hand, excited voices claiming, "See, we told you. It was a disaster and it can happen here"; and on the other, by soothing murmurings assuring us that "It was of no real consequence."

Zhores Medvedev may have been unintentionally exaggerating when he claimed hundreds of deaths and thousands of injuries. Only the Soviet government knows for certain. And it is possible even they don't have a clear count.

More significant in the long run than the number killed or injured is that even with close scientific and governmental supervision, at least one, probably two, and possibly more disasters did occur. Some injury and loss of life was sustained. And a large area of land was devastated and will not return to productivity in the lifetime of anyone living today.

No matter how careful we are and how limited we have made the risks, we are still in some peril. Not more than we can afford with rational preplanning and individual concern, but more, certainly, than the pro-atomic energy forces concede we face.

The Soviet disaster or disasters show what a tightly centralized government can do if it wishes to expend enough time and energy on a coverup. While the intelligence forces of the world store away the knowledge of an atomic disaster, the people remain uninformed.

Fortunately, it would be far more difficult to achieve this same effect in the United States—but probably not much harder to have an incident of an even bigger magnitude.

No real thought, at the time of the Ural accident appears to have been given to even simple matters such as evacuation routes and temporary shelter for homeless people. Realistically, we must conclude little more has been done today.

43

Here in the U.S., emergency plans are a part of every application for the license to construct an atomic power station. And these same contingency programs exist around each of our plutonium manufacturing and weapons production centers, as well as every other site where atomic materials are processed in any great amounts.

A look at one of these disaster plans is interesting. Most leave a lot to be desired from both the standpoint of operational feasibility and cooperation with state and local governments. A lot more work needs to be done in this field.

■ One of the hardest things to account for in all estimates of nuclear plant failures is not the problems caused by internal equipment break down but those which might stem from an external source.

As we have seen from the Soviet example, lack of planning makes any mishap all the worse. The longer it takes to mobilize an effective countermeasure, the greater the losses.

But what about the eventuality for which no planning is possible? The one which occurs not because of some electrical or mechanical malfunction, but stems from a direct, deliberate, premeditated attempt to sabotage an installation.

Such an action would have taken all safety contingencies into consideration. Fail-safe mechanisms would be bypassed or double-faulted so as to surmount them. Deliberate sabotage is unforseen as far as the moment of its occurrence.

Who would be insane enough to try and sabotage an atomic installation? From published reports there are lots of people.

As difficult as this is to imagine, 11 separate attacks have been made on nuclear power stations alone. They happened in various places around the world. Six took place right here in the U.S. Additionally, 8 threats, with serious overtones, have been recorded, and 13 cases of vandalism or deliberate sabotage have been mentioned publicly. And none of these included the 15 incidents in which vital security breaches have taken place, or the 11 U.S. companies fined for noncompliance with nuclear security regulations.

All together, there have been 64 notable incidents worldwide since 1966. And the brunt of these occurrences has taken place after 1974.

Surprisingly enough, few really big news stories have come

out of all this, which leads the antinuclear forces to suspect some kind of control of the press. But the answer is simpler.

We just don't want to hear about it. In a way, we are concerned, but it is easier to relegate our occasional worries to a narrow channel in the backs of our minds, where they join with blasé attitudes toward other potential dangers in our society.

A close look at some of the incidents which have occurred, however, will show the potential seriousness of this not too often discussed problem.

The number of attempts seems clearly to be increasing, as proliferation of atomic energy continues and knowledge of its handling and various potentials becomes better understood by nonscientists.

The government recognizes this problem and has asked for stricter security controls at all nuclear power plants and other installations. According to a report issued by Morris K. Udall, D-Ariz., and Paul E. Tsongas, D-Mass., on behalf of a House subcommittee, the Nuclear Regulatory Commission (NRC) acknowledged the security deficiencies, but many had not been corrected.

According to Udall and Tsongas, the Commission was especially troubled about the possibility of attacks by three or more "well-armed, well-trained persons who might possess inside knowledge or assistance."

They further stated, "The Commission believes that such groups might possess explosives, machine guns, anti-tank weapons and helicopters." What makes this possibility so alarming is the sum total of defense and fire power possessed by the average nuclear installation consists of only two guards "armed with .38 caliber revolvers and shotguns."

A letter dated July 20, 1976, from NRC Chairman Marcus A. Rowden revealed security deficiencies at every one of the fifteen installations handling plutonium or highly enriched uranium. Udall and Tsongas, following up on this information, discovered seven of the fifteen installations continued to have security breaches "despite the earlier investigation."

In a more recent situation, a reporter, Joseph Albright, posing as a contractor, sent the Air Force $5.30 and received through the mail the detailed plans of a nuclear weapons depot. He then personally visited two sites, where he was shown a guardhouse and told the routes and approximate timetable for trucks carrying atomic bombs to waiting aircraft. The only identification required of Albright was a driver's license and a few credit cards. Apparently no one checked to see if he was really a contractor. As a result, Donald R. Cotter, a Pentagon weapons chief, is reported to have ordered an immediate investigation of Defense Department procedures.

Partially as a result of the House subcommittee report and partially due to a growing concern of several members of the NRC, strict new rules were proposed to safeguard all atomic installations. These requirements, set forth early in 1977, recognize the need for greater care both at the power production and nuclear manufacturing sites and while materials are in transit.

News accounts of the revised proposals made it clear the NRC had given some thought to direct terrorist or guerrilla intervention by armed force into the premises of the many civilian-operated nuclear facilities.

Owners and operators of these plants were ordered to submit new and more far-reaching plans for increasing security and protection of the vital equipment and controls.

The NRC's action set off a mad scramble for the development of the required programs within the 90-day limit granted by the committee. According to one utility executive, the speed demanded by the body's action "did not leave much time to really think about all the questions."

By the week of June 13, 1977, 28 United States utility companies had turned in security programs for their 44 nuclear reactor power stations. They were granted until August 1977 for NRC's approval and the complete installation of the various devices they felt they would need to defend themselves against a determined effort by a small but heavily armed and knowledgeable force. The same date was set for the addition of manpower outlined in the various proposals.

The cost of this added security is high. Estimates indicate the tighter requirements have opened up a $100 million market in 1978 alone, ranging from closed-circuit surveillance TV to special fences with barrier alarms. All told, about 90 percent of the atomic power plants in the U.S. will need added security-related products to comply with the new regulations.

James R. Miller, the NRC assistant director for reactor

safeguards, estimates a plant's investment will be about $2.5 million for the initial setup, and another $1.5 million annually for maintenance. Compared with security costs in other industries, this is a reasonable amount.

In addition to increasing the number of guards at each plant to a minimum of ten and requiring two hundred hours of specialized training for every man, bullet-resistant glass will have to be installed in selected areas. Tamper-proof locking systems, for which no key can be duplicated, perimeter intrusion alarms, massive fencing, extra command posts, and remote-area television surveillance are also part of the new package.

But security consists of more than guards and gadgets. It is an attitude. And overall, judging from the fact that eleven firms have been fined since June 1974 for noncompliance with security regulations, this necessary attitude is lacking.

As difficult as this may be to accept, the reasons for it are clear. First, the utility companies who operate most of the nuclear powered generators have long been associated with the production of electricity by more conventional means. The old coal, gas, or oil-burning plants have never really been a target for terrorist groups here in the U.S. Aside from a few bizarre and relatively unthreatening incidents, the business of producing electricity has been isolated from the mainstream of violence. In short, due to past experience, management attitudes are not directed toward this problem.

The second major reason for a lack of concern stems from the rather cursory treatment given to this matter by the U.S. Government and regulatory bodies.

It was not until July 18, 1973, that the American National Standards Institute issued ANSI Standard N 18.17 entitled "Industrial Security for Nuclear Power Plants." This document established the first really detailed measures for protection of the reactor installation.

Later, in the November 13, 1974, issue of the *Federal Register*, the then-active Atomic Energy Commission (AEC) proposed rules for the protection and physical security of nuclear power plants and materials.

These first steps were very general in nature, indicating the necessity and advisability of security, but few specific requirements were established.

This was the state of things until recently. In the time between the publication of the original ANSI Standard and the new NRC requirements, the AEC, through on-site visits and a review of security systems, was able to begin development of tighter controls. Ample material was available for detailing these safeguards, but the AEC was phased out of existence and the NRC became heir to its work in this area.

No clear indications are available as to the effect these measures will have on armed attempts to take over atomic installations. But judging from the thoroughness of the recent rules, especially as compared with the almost nonexistent prior regulations, a terrorist organization will have to expend a lot more effort to capture a nuclear power station.

Some of the NRC's action in the matter of security may have been spurred by a series of reports from the GAO which was highly critical of the previous security and requirements.

The GAO, acting in an investigative mode, sent its auditors to one installation where they gained access to vital areas by picking the door locks with pieces of discarded wire and an old screwdriver.

They also learned, as their study progressed, of a plant protected only by an eight-foot wire fence, of incidents in which full-time guards had been given less than four hours' actual training, and of situations where the security force could not operate the TV scanning equipment or find the main control room.

The GAO recommendation eventually went so far as to suggest giving N-plant police the right to "shoot to kill" in some conditions. This brings forth the frightening prospect of a private army whose ranks are filled with men granted privileges above the law.

In acknowledgement of the utilities' past faults, Phillip J. Cherico, director of security for the New York State Power Authority, said "Security has not had the priority it should have had, particularly among private utilities. It does not," he continued, "contribute to the generation of either power or profits."

Estimates indicate the newly revised NRC rules will require only a small-percentage increase in the plant operating budget for full compliance. One source indicated about 2 percent of the total annual expenditures would be needed for all security measures. Another said the new rules would increase an average consumer bill by less than a nickel a month.

Some resistance to full compliance, based mainly on isolated points, is justified. The regulation, for example, which specifies each guard be equipped with an automatic weapon, because he might end up fighting terrorists with similar armament, is a real quandary. In addition to running afoul of numerous state and local laws restricting the use and possession of these guns, a great deal of experience is required in their use to avoid shooting hails of hot lead into crowds or traffic on nearby roads.

Another difficulty is the stringency of the search requirements. Every person and vehicle entering a power site will be subjected to a very thorough going-over. Frisking union-member employees is almost certain to bring up labor trouble, and the time required will lower productivity. In addition to reactions by civil rights groups who see this infringement as the beginning of a police state, the unions want to know if the search procedure will be on company or employee time.

All this adds up to a long and protracted fight to implement the needed security standards and maintain a semblance of status quo in the efficiency of the work force employed in the installations.

Some of this problem will be eased by available technology. In addition to the expected TV cameras and electrical barriers, the use of reliable walk-through metal detectors and gas sensors to spot guns and explosives, will allow for quicker screening.

But quick or slow, no form of spot check will stop incidents such as the GAO lockpicking or the arrival, in a German nuclear plant, of a man in the main office area armed with a bazooka-type antitank rocket launcher. Even though the last incident was done only as a protest, the presence of a portable weapon capable of causing sufficient damage to wreck the pile is intolerable if we are to maintain even reasonable security in our stations. And quick employee checks will not prevent unauthorized entry. The addition of trained security forces on alert patrol provides the only deterrent to this threat.

Critics of nuclear energy in general have enjoyed a boost by the revelations of the various studies of atomic power plant security. And again, the proponents of nuclear power have come forward to down-play the talk of holocaust.

Responsible middle-grounders view the problem as serious but not disastrous. They are glad to see more strict security regulations, expect difficulty installing the controls as a practical matter, and look forward to the day when every facility will be adequately policed by its own internal staff.

They also recognize the inherent danger in complacency. No security, no matter how well formulated, will ever be enough to stop a serious, well-organized, concerted attack. The only possible hope in such an instance is to hold on until help arrives.

During the Viet Nam confrontation this point was proved beyond all hope of rebuttal.

A team of specially trained Americans was flown to Viet Nam with the mission of destroying the only nuclear reactor in South Vietnamese control, to prevent the Viet Cong from obtaining radioactive materials which could be fabricated into weaponry.

The task was easily completed in record time. One official source holds the reactor itself was destroyed. Others maintain only the nuclear fuel was removed from the body of the pile.

This operation required specialized equipment, trained individuals, and support teams to evacuate the group after the attack was complete.

None of these requirements is more than could be met by any one of many terrorist groups in existence today. There are, throughout the world, ample targets for this kind of operation. And probably the only reason one has not yet been attempted is due to the attitudes of the insurgent groups toward atomic energy. The argument that terrorists want to create only the kind of havoc which produces a large audience is a weak one. If the group is after absolute negotiating power, there is probably no more effective way to achieve it than attainment of some form of nuclear weapon. And as far as getting attention, modern mass media will see to it the event, whenever and wherever it occurs, will be brought to the public's attention, unless governmental security is clamped tightly in place.

But there is yet another threat to the basic security of any atomic installation.

The desperate or deranged individual, acting out of a strong conviction or a moment of panic, could place himself in control of the lives of thousands of people.

It sounds farfetched. The pro-atomic forces have scoffed at the idea, and once more, the antagonists have cherished it.

Yet such an event really took place. The participants successfully held an atomic installation for ransom and caused the evacuation of the most prestigious nuclear operation in the United States. They did it in a surprising manner, using a technique which could be duplicated by other groups.

The drama made headlines for a single day, back in the early 1970s, and is now largely forgotten. It shouldn't be, because it demonstrated that what possibly can happen, will, and in a most unlikely fashion.

Three men were involved. They had no special training, no real knowledge of nuclear facilities, and, fortunately, no strong purpose other than the attainment of a large sum of money.

The security measures utilized today, including the X-ray machines, TV systems, electronic sensors, metal detectors, and armed guards did not stop them.

They came in on the blind side.

From the air.

■ 5 ■

■ It was Friday, November 10, 1972. 10:00 P.M. CST. The guard arched his back wearily, wiggled in his straight-backed chair and settled himself again. The coffee in the plastic-coated paper cup was lukewarm and he sipped it more from habit than need. Pale with cream and sugar, it tasted sweetly bland. He made a little face as he replaced it onto the Formica desktop.

The images on the television screens in front of him cast a greenish flicker as they changed, one at a time, to a pre-set cadence. Every area of the main reactor center of the AEC's nuclear research installation at Oak Ridge, Tennessee, came into view, lingered for a few seconds, then vanished, only to be replaced by more pictures of asphalt paving and concrete walls.

The guard shifted again. Lulled by the constant replay on the 18-inch tubes, he allowed his gaze to wander to the large clock on the wall and shook his head as he realized the big hand had moved less than an inch since his last bored glance.

The routine was endlessly the same. He'd be alone for another few minutes, then be joined by one of the roving guards who patrolled the area. They'd talk, disconsolately, for a while, then he'd be alone again, monitoring the entire complex over the closed circuit.

Something moved in one corner of the screen, but the picture changed to the view from another camera before he could detect what it was. Lackadaisically, he pressed the proper button, snapping the previous scene back onto the tube.

The uniformed man relaxed. Nothing. A maintenance truck parked next to one of the service entrances had started up just as he'd focused his attention on the picture. He watched the truck drive away slowly, taillights gleaming blackly on the tube where the brilliance caused the electronic picture to drop out, then he returned the set to automatic scan.

53

In the months he'd been on duty in this installation nothing varied the routine. An occasional false alarm was a welcome event. He sighed and reared back in the stiff-legged chair. His bulk caused it to squeak loudly in the silence of the coldly efficient room, and he eased up, only inches from toppling over. Another day, another dollar. His gaze returned to the clock, and the big hand, which had moved only slightly since his last examination, held his eyes as his hand reached unerringly for the half-cold, almost empty cup. The scenes on the TV tubes continued to flicker and change.

■ High above the dark, rolling Alabama countryside, Southern Airways flight 49 ripped its way through the blackening sky, on schedule out of Birmingham and due to arrive in Miami after an intermediate stop in Montgomery. Things seemed normal. The ground below, dotted by individual lights from scattered farmhouses sparkling in the growing dusk, looked much like the star-strewn sky above.

The jet, which can seat seventy-five, was about one-third full and the two stewardesses were moving through the cabin caring for their charges. The passengers had just settled down from the nervousness brought on by takeoff, and were ordering drinks. They were on course and on time. Tension drained from the cabin.

Three of the passengers, however, were still uptight. One of them nervously kicked the small satchel he'd slipped under the seat in front of him, then leaned back and tried to relax. At twenty-one, Melvin Cale had been in a number of tight places. Since his escape from the minimum-security rehabilitation program to which he'd been sentenced in 1971, he'd been on the move, avoiding notice. Now, on board with his half brother, Louis Moore, and Moore's inseparable companion, Henry Jackson, both of whom were also fugitives, he was about to make world headlines.

Moore and Jackson had been through a lot together. They'd had a restaurant, "Lou and Smooth's Soul Palace," and they'd done okay for a while. But then things turned sour and they'd given it up. Wasn't any money in the two-bit place anyway. And it sure took money, because the real color of segregation was green, baby, not black.

Karen Chambers, one of the two stewardesses, passed in the

aisle again, and Louis Moore followed her with his eyes. His mind wandered, then centered on the thought of the Detroit cops and their rape charges. Hell, he'd gone down to the station with Smooth Henry Jackson to lend a little moral support. They went everywhere together anyway. Then that sassy little lady cop knocked on 'em, and the next thing, they were charged with rape. Them. They'd made speeches before the Detroit City Council and were on their way inside the system to be somebody.

The hot recollection of the scene at his home with his wife crying and the two kids looking scared flashed into his reverie. Hell, he'd make it up to them after this little number.

Smooth Henry nudged him lightly, and when he looked, was nodding his head. One thing about Smooth Henry. He didn't have no bad home life. His wife was an ex-Playboy Club Bunny and did go-go dancing when her old man needed bread. He knew Henry's tastes ran to cold liquor and hot women, so there was little question in his mind what he'd do with his split of the ransom.

Taking the money wasn't stealing either. The money was coming from the same city that once had the gall to offer them $25 in settlement of their $4 million suit for police brutality.

Melvin Cale sat alone, apart from his two companions across the aisle. It had seemed natural, when boarding, to let the other two men sit together. They did everything together.

Cale twisted his body to assume a more comfortable position. He was nervous and it was hard to keep from fidgeting a little. Lost in thought, he was paying no attention to his surroundings, and he gave an involuntary little jump at the sound of the stewardess's voice.

"Can I get you anything, sir?"

Cale shook his head. "No, nothing." He looked across at the other two, who were staring back at him. Not time yet. He tried again to relax.

It was hard to do. He was uptight a lot since he'd fled the jail-release project in Nashville. He'd been in trouble with the law much of his life and the last episode had ended with a five-year sentence for larceny. When the opportunity had come to run, he did.

The droning of the aircraft was soothing, and he looked across the narrow aisle at the other two. They were shrewd. Real dangerous cats from up north in Detroit. And running from the law

55

too, no matter what they said. It was their scheme, really. He'd been called in to help and had added some ideas, but it was their deal from the start.

Henry Jackson looked over toward his friend, then at Cale, and winked. He pointed to his wristwatch. Not long now.

More out of nervousness than anything else, he leaned forward, and unzipping the canvas flight bag which rested on the floor, peered inside at the cold, shiny, chrome-plated steel barrel of a revolver. Satisfied, he rezipped the bag and straightened in his chair.

Not long now. They'd discussed it and wanted to give things a few minutes after takeoff to settle down. Time to get the airplane away from the field. Out where no one could interfere.

Melvin Cale got up and out into the aisle as he saw Jackson start forward. Lou Moore remained in his seat but was cradling something against his stomach.

In the forward part of the cabin, Donna Holman, the second stewardess, was pouring coffee into a cup. The captain always liked to have it served immediately after takeoff, and she knew a happy captain meant a happy crew.

Turning toward the flight deck, she saw the stocky, round-faced man coming up the aisle. His Afro haircut and mustache marked him for a Northerner. She paid little attention to him as her mind was more on the task of not spilling any of the hot coffee.

The man approached her before she could move, talking excitedly. She tried to turn away but he slipped his arm around her, and she felt the cold metal barrel of the fancy chrome-plated .38-caliber Smith and Wesson he carried dig into her neck. She resisted for a second, then slumped.

Midway back in the cabin, Louis Moore, seeing the stewardess safely captured, stood up in his seat and began waving a black World War II Luger pistol in the air.

In the rear, moving swiftly, Melvin Cale, who had started for the back as Jackson moved forward, grabbed the other hostess, Karen Chambers, and shoved his cheap .22-caliber Saturday Night Special into her so hard she winced in pain.

Jackson's voice was loud in the cabin, overcoming the noise of the engines and the rushing air outside. "Nobody move. We're taking over. And let's not have any heroes. Because they'll be dead heroes."

56

The frightened passengers reacted in a variety of ways. A man started to rise, then, thinking better of it, settled back in his seat under the watchful muzzle of Lou Moore's automatic. A woman seated next to her teenaged son grabbed his arm, and speaking in a low voice, tried to reassure him.

Jackson, apparently familiar with the routine of the plane's crew, moved rapidly. With his strength and threatening pistol, he forced the stewardess to the narrow door leading to the flight deck.

Following standing airline orders, she did not resist him. At the door he spoke to her urgently and motioned with the gun to the lock. They stood for a moment in conversation, then, with a look of resignation, she took out her key, and moving with great care, opened it.

On the flight deck, Captain William R. Haas, a veteran pilot, and Billy Johnson, his second officer, were at the controls. Haas, forty-three, was a calm man with great flying experience. He was on flight 49 as a favor to another pilot. It was his day off, but he had shifted his schedule to accommodate a friend who needed the free time.

The on-course climb out had gone according to plan, and the two flyers had no inkling of the action behind them in the passenger area. They had just finished making their normal course corrections during the leveling-off procedures when the door burst open.

Turning, expecting to see the stewardess with a cup of hot coffee, almost a ritual for Haas when in the air, he was surprised to find a wild-eyed man holding the woman captive and waving a shiny pistol. An icy calmness descended over the captain, and he faced the situation squarely.

The girl, Jackson's arm still around her neck, spoke quickly. "This is a hijacking. He's not kidding."

Haas looked intently at the man and realized the seriousness of the situation. He held his voice down, maintaining a placid exterior. "Keep calm. We don't want anyone hurt. We'll obey your instructions."

Jackson, confused by the array of gauges and controls and made nervous by the tight confinement of the small space, was wary.

Haas's demeanor, however, put him at ease. Sensing the

57

captain was telling the truth, he released the stewardess and sent her back into the main part of the plane where Lou Moore was standing on one of the seats, waving his pistol at one after another of the passengers.

Jackson waited until the girl was gone before speaking again. "Get this plane to Detroit. We want ten million dollars from the city of Detroit or we're gonna start killing people on this plane."

Captain Haas, a native of Moscow, Tennessee, and the copilot, Billy Johnson, who, when not flying, served as mayor of College City, Arkansas, were wise, experienced men. They had the full responsibility of the crew and passengers' lives in their hands and were well trained in the basic company policy dealing with hijack attempts. With Jackson, Moore, and Cale clearly in control, their orders were to cooperate until they were rescued or released.

Haas, trying to lead Jackson into a conversation, outlined the problems of changing course for Detroit. First, there was the FAA and a flight plan. They would have to be cleared through a corridor of air on a course which would take them there. Then there was the matter of insufficient fuel. They would have to stop somewhere to refill their tanks.

He and Jackson spoke for several minutes. Haas outlined the radio procedures necessary to put out the ransom demand, get clearance, and the number of airports where they might find the right kind of fuel for the DC-9. Jackson moved back into the main cabin, and keeping an eye on the two officers, motioned to Lou Moore, who came forward. They talked briefly, then Moore walked back to the waiting Cale to explain their plan. Jackson returned to the flight deck.

He and the pilot exchanged a few more words before a decision was made. "Then get gas at Jackson," the gun-wielding man said. "And nobody and nothing near this plane except a gas truck and driver in his undershorts."

Haas went back to the radio and the plane continued its way through the night.

■ The guard shift had changed at Oak Ridge Atomic Laboratories and the new men had taken the positions vacated by the retiring group. It was almost Saturday, November 11, and the endless displays of television surveillance pictures continued their regular

58

mindless pattern across the battery of cathode-ray tubes. The routine of the day remained unchanged.

■ The DC-9 had been rerouted from its original course and given an immediate clearance to Jackson, Mississippi. Word went out as soon as the hijacked condition of flight 49 became known, and agents of the FBI met with Southern Airways representatives within an hour after receiving notice.

The demands of the three were discussed, as was the impossibility of attaining large sums of cash at 7:00 in the evening. A long chain of events was set into motion and 24 hours would pass before it would all end.

Conversation between the pilot and the armed men was terse. Knowing it might take some time to gather up the demanded money, the three planned to keep the aircraft aloft as much as possible where interference would be far less likely.

Shortly before 8:00 on Friday night, the plane rolled to a halt on one of the runways at Jackson. The refueling operation took a minute to start, and one of the two men stationed in the main cabin came forward and passed the hijacker in the flight deck a heavy object. Pulling the pin, Henry Jackson held a hand grenade next to the ear of Captain Haas. The pressure of his fingers on the narrow safety handle kept the primed bomb from exploding. His meaning was clear. There were to be no tricks.

The people inside the aircraft stayed away from the windows, and two of the three bandits remained on watch constantly as the truck approached to hook up its fire-preventive ground-static discharge wires. The hand holding the grenade remained near Haas's head and Jackson's eyes watched the operation with intense concentration. There was little conversation.

On the ground the authorities had rushed two agents to the scene, but they were unable to approach the aircraft. Identification of the three hijackers was made, however, and on the off chance some good might be done, Cale's wife and daughters were sought so they could be brought to the Knoxville airport to talk to him over the radio to dissuade him from killing the people on board.

In the main cabin, all male passengers had been ordered to strip down to their underwear in an attempt to keep them under control by embarrassing them.

After about an hour on the ground, during which time the fuel truck had topped the big airplane's tanks and driven away, Jackson grew nervous. The increasing number of radio transmissions and his unfamiliarity with speaking over the microphone, along with the tension created by being exposed in a static position, caused him to react.

"Take off." His voice showed his nervousness and Haas responded quickly. Tower and runway control permission had already been granted and he spurred the ship into immediate motion. They rolled quickly down the long runway and were airborne in minutes.

Once aloft the three men seemed more relaxed. Jackson, working with exaggerated care, replaced the pin in the grenade and passed it back to Lou Moore. They were on their way again. An air route was chosen and air traffic control cleared a flight corridor for the plane toward Detroit.

The three were joking and singing but continued their close scrutiny of the crew. One of the male passengers was looking ill, and the stewardesses saw to his needs as best they could.

Several radio messages were exchanged as the plane flew northward. Progress reports on the arrangements for the ransom were made as if the full amount were being raised.

A major mobilization was going into effect. The FBI had alerted its field agents over a large zone, and the U.S. Air Force offered the use of military jets to assist in the hunt.

Once in the Detroit area, Jackson and Moore had the pilot place the ship into a circular holding pattern. It was like having their own special scenic cruiser as they gazed down at their city from a majestic height. They had arrived back in style and were safe from the police miles below.

A meeting of law enforcement groups, representatives from Southern Airways, and members of the Federal Aviation Administration (FAA) had resulted in a plan. The situation while the plane was in the air was hopeless. Short of shooting it down, there was no way to abort the flight. But once they had the craft on the ground, even in the open spaces of an airport, there was a chance. One or more of the men might expose themselves to specially trained and selected sharpshooters. Or a team might gain entry by a ruse. As long as the ransom collection could be

delayed, and the hijackers continued to believe there was a possibility they would be paid, things could be held in suspense.

The three on board, still keeping the passengers in a state of
terror, had ordered the liquor locker opened. Inside, the small
bottles of Scotch, bourbon, vodka, and gin were rummaged
though, and the hijackers began to drink.

In an aisle seat, eighty-three-year old Alvin Fortson, a farmer
who had been showing signs of breathing difficulty, now seemed
to be having a mild heart attack. One of the passengers requested
and received permission to come to his aid and helped him lie
down across three seats. After loosening his tie, belt, and shoes,
the man massaged Fortson's legs to help circulation. But his
condition worsened. Finally, the stewardesses asked if they
might give him oxygen from the emergency equipment, and Cale
agreed.

Radio messages continued to be exchanged and the three
armed men talked among themselves. They brought up different
ideas to confuse the authorities about the number of hijackers
actually on board. Unknown to them, a previously agreed upon
emergency code was being used to keep the ground stations
posted. Southern Airways representatives continued to assure the
men work was progressing on the roundup of the required ten
million, but were quick to add it was taking time.

The scene in the control area of the jet was tense as they
circled the city. The three bandits, taking turns standing guard
and resting, were an ominous presence. The captain and his second officer, consuming cup after cup of steaming coffee, were
feeling the combined strain of the irregular in-flight operations
and the hours they had spent at the controls. The two had done
everything in their limited power to ensure the passengers of their
individual safety, but the sight of the ship's officers under gun-
point was demoralizing to everyone on board.

As soon as the basic desires of the three terrorists had been
clearly understood by the FBI, they developed a plan. Money
would be used as a lure to bring the big airplane to earth. And
since the ransom had been demanded of the city of Detroit, it
seemed an ideal place to try the trick.

Detroit Mayor Roman Gribbs was contacted and told of the
men's requirements. Surprised and incredulous, he replied, "Are

you kidding? That much money? Why would they want it from the city?" When he was given a quick outline of Henry Jackson's and Lou Moore's previous records, Gribbs understood and acted. After consultation with Southern Airways spokesmen, he moved with great effectiveness. A midnight emergency session of the city council was convened and a loan of $500,000 was immediately approved.

Gribbs's decisiveness and action provided the FBI with the bargaining tool they needed. They now had a large amount of cash and could try and trade with the three hijackers.

But as rapidly as things had been done, they had taken valuable time.

After circling for almost two hours, the DC-9 was again low on fuel. And the three desperate men were running low on patience.

Cleveland was chosen as a satisfactory landing site for refueling. Without much warning, the DC-9 altered course and the pilot called for a new flight corridor.

The authorities, sensing an opportunity, quickly rushed a squad to the Cleveland airport in an effort to stop the takeoff.

By radio, the three men, still trying to confuse the issue of how many were really in their group, asked for food, additional drinks, and ten parachutes. These were to be put on the aircraft while the refueling operation was in progress.

The brief flight from Detroit to Cleveland hit a new high in tension. Fortson, the eighty-three-year-old farmer with the lung and heart problems, had taken a turn for the worse. In spite of the attentions of one of the passengers and the stewardesses, it seemed he might be dying. Oxygen was given to him on a continual basis. Melvin Cale, gun in hand, in an act of either bravado or attempted humor, would grip the old man's wrist and study his watch, as if timing his heart rate. Then he would smile and nod sagely. He apparently liked playing doctor, but obviously didn't know how to take a pulse reading.

The special agents reached the Cleveland airport ahead of flight 49 and held a quick meeting to go over the various attack possibilities.

A rapidly mounted plan went into effect as the large aircraft shut down its engines at the end of one of the concrete runup areas. While the refueling took place, a specially trained team tried to make visual contact with the hijackers by bringing the food and water on board. The sight of the armed hand grenades dissuaded this group from any action, but they tried to hold their attention while FBI marksmen moved up within shooting range.

The ploy failed. One of the terrorists, posted as a lookout, saw the men advancing across the open area of the runway about a hundred feet away. The hijackers' response was instantaneous and directed not toward the crew and passengers held captive, but against the attacking agents.

Calling from one of the small windows in the flight cabin, a single voice was heard over the noise of the refueling truck: "Get the hell back or we'll throw a grenade."

Convinced it was no idle threat, the sharpshooters retreated. At this point, things relaxed a little. With replenished fuel, food, water, and ten parachutes, the men felt better. And aside from the single, aborted try to recapture the aircraft, no resistance was in evidence. It was apparent they were going to be very hard to stop.

Less than an hour after landing, the DC-9 was in the air again.

Jackson, after discussing the matter with his two associates, gave Haas the new destination: Toronto, Canada. The radio once again was busy with messages.

Confident of their newfound bargaining ability, the three became more insistent about their desire for immediate payment of the ransom. Jackson, speaking to Haas, sounded more intense than ever. "Tell 'em," he said in a threatening tone, "that they goddam better have the money there."

What had up to this point been a national problem now reached the status of an international incident.

■ Communications between Canadian law enforcement officials and U.S. forces were easily established. Air hijacking was a problem of concern to all countries, and the Canadians were glad to help out.

A long series of conversations ensued between the FBI and the bandits. Montreal was agreed upon as the landing site and the place the ransom would be paid.

Time was beginning to favor the police. The plane had been in the air over twelve hours and even with food and water resupplied, the flying officers had to be getting weary. This also brought on a renewed concern for the passengers.

A total of $500,000 in small bills, the amount Southern Airways had borrowed from the city of Detroit, was delivered in a special courier plane. A Canadian police officer with an exceptional record of service offered to strip himself to the skin and present the money in person so the three on board could see there was no trickery. The proposal was made and refused. Negotiations were continued as the plane once again touched down and was refueled.

There was no chance the three men would settle for a mere $500,000. Their original demand for a full $10 million was renewed, but his time in a tone which made it clear the high spirits the hijackers had when leaving Cleveland were gone. Threats against the lives of the passengers and crew were renewed.

In about fifty minutes the plane was in the air again. Heated exchanges came over the radio. There was no question in the minds of the law enforcement officers the matter was taking a nasty turn. But just how nasty nobody really knew.

The three hijackers were at a loss. It was obvious to them the flight officers, who had now been at the controls for more than twelve hours with only minor breaks allowed for trips to the lavatory, were approaching their limits of endurance. And the passengers, fatigued by the constantly oppressive fear, were becoming inured to the danger of the situation. Somehow, the threat the three held over their lives, as well as the valuable airplane, had not been enough to get action. Something far more drastic was needed.

Jackson, Moore, and Cale conferred briefly, keeping a careful lookout for any sudden activity by the captain or the crew.

The weather outside the DC-9 was as wild as the scene inside. Wind, rain, and snow swept the plane and caused it to bounce like a cork in rippling water.

No one is certain which of the three came up with the idea, but Cale had lived in the Oak Ridge area and knew of the fear, and thus the powerful potential for negotiations, attached to the atomic reactors installed at the nuclear facilities. The instigator is unimportant, however, because all three men agreed to the following course of action.

Jackson's voice over the radio was firm. The reaction in the aircraft control tower in Toronto was one of stunned silence. "We're tired of all this bull. No more foolin' around. We're taking this fucker to Oak Ridge and dive it into a nuclear reactor."

From brief conversation between the startled tower personnel and the now-committed men, two things emerged. First, Jackson, Moore, and Cale were sincere. If they didn't get the $10 million, they would crash the plane into the Oak Ridge atomic installation. Second, they gave a deadline, after conferring with the captain about flight time, of 8:00 A.M. for arrival in the area and a 12:00 noon final ultimatum point. Either they got their money by then or they would deliberately crash into one of the nuclear reactors. What had been a threat to the lives of twenty-seven passengers and four crew members now became a promise of potential national disaster with the well-being of innumerable people in the balance.

Top-level meetings were convened. The Oak Ridge staff was alerted to the danger, and new, harsher measures were considered for trying to capture the three hijackers. Through the early hours of the morning various experts and field agents were consulted, while frequent radio messages continued to promise quick delivery of the ransom.

On schedule and tracked all the way by FAA radar, the DC-9 arrived in the Oak Ridge area and began another of its unending circles.

When the call came through from Canada to the Oak Ridge guardroom, the man on duty at first thought the story he was hearing was a hoax. Then, slowly, his disbelief changed to horror. He contacted his superiors, and orders went out immediately for a meeting of all key personnel to consider possible emergency action.

The ultimate threat had been made. And there was no defense possible. If the three men on the plane could overcome the pilot's resistance to crashing directly into one of the reactors—and this seemed highly likely—no one was certain how much damage would result.

But everyone agreed on one thing. If the crash actually occurred, the impact could rupture the protective shell and there would be a release of radioactivity into the environment. That was the least of the projections. The worst indicated the possibility of a full-core meltdown and the issuing forth of enough contamination to make the day one which would long linger in the less pleasant memories of man.

The evacuation of the Oak Ridge laboratories did not extend to the nearby communities. Government spokesmen, acting on the least serious of the damage projections, issued a calm statement to all news media indicating there would be little or no release of radioactivity if the three men should succeed in their plan.

Great controversy was eventually to arise out of this statement and the actions of the concerned officials of the Atomic Energy Commission.

After circling for two hours, waiting for the final 12:00 noon deadline, the three hijackers were once again told by Captain Haas that they were low on fuel. Lexington, Kentucky, was chosen as a stop, and they landed. The fill-up was made without incident, and in less than 30 minutes the ship was airborne again.

The attitudes of the people aboard flight 49 had gone from calm aquiesence to near panic. Everyone grew progressively more tense as the three made their plans clear. No one present doubted their seriousness.

Twelve noon was approaching rapidly as the relentless circling began again.

The Air Force offered planes to the FBI and several fighters were sent aloft. They were under no specific orders, but could, if called in time, make a pass or two in hopes of shooting the DC-9 down before it could actually dive into one of the atomic structures. The authorities, wrestling with the morality of killing all thirty-one on board and possibly sparing the lives of a larger but unknown number, were close to action but not yet ready to take the final step and order an attack by the silver jets.

Still convinced money was the answer, a continuous battery of assurances was broadcast. The money was on its way. Hold on. Don't do anything rash.

But the three men on board were in a highly agitated state. Lou Moore, talking to the passengers, revealed his position clearly. The situation was getting hopeless. Death might be the only way out. And a far better alternative to prison. "I was born to die," he said, "and if I have to take all of you with me, that's all right with me."

Haas, concentrating his tired mind on precision flying,

66

realized a climax of sorts was being reached. All through the long night he had done his best to cheer the weary passengers and encourage them. But things were looking more hopeless.

Finally at 12:00 noon, with no real progress made on a delivery place and time for the ransom, Jackson acted. He pointed his chrome-plated .38 at Haas and said in a low serious tone, "Dive it into a reactor."

Haas thought as quickly as his fatigue would allow. "I can't," he replied slowly. "There's an overcast beneath us and there's no way to tell where the reactors are."

He was convincing, but Jackson was undeterred. His mind was made up. Then, the first break in the situation occurred. Haas told Jackson, and apparently was able to convince him by his discussions over the radio, that there had been a mixup in the times due to crossing from the Eastern to the Central Standard Time zones. The money was coming, but everyone thought they had until 12:00 Central, not 12:00 Eastern Time.

The ruse worked. Haas gained another hour, which was to become the critical factor.

Jackson, with some of the tension off for a moment, dictated a radio message which the copilot, Johnson, sent to the area ground stations. "The 1:00 o'clock deadline isn't far off. We must have the money, seven bullet-proof vests, helmets, and a document from President Nixon granting us the ten million. We must have stimulants for the crew, food, a six-pack of Pabst Blue Ribbon Beer, coffee, water, and cigarettes for the passengers. If the conditions are not met, we take this thing into an atomic energy plant."

Down below, through the broken overcast, the sound of the circling plane could clearly be heard as it passed over the now almost deserted Oak Ridge facility. The evacuation of non-essential personnel from the site had gone according to plan and nothing stirred on the usually busy streets. In the nearby towns, the natives went about their normal routines. There was some extra traffic out of the area, but by and large people maintained their regular schedules. Many did not know until later about the seriousness of the threat.

The clock ticked the time slowly away, and the interval between noon and 1:00 P.M. lessened. The radio crackled with progress reports and promises, but the three hijackers seemed to be in another world. They were going to get their money or die.

And if the alternative was to die, they would take a lot of others with them. Haas had done his best. He'd worked against hope to gain the last possible time extension, and now it was running out.

More Air Force pilots were alerted. The shoot-to-kill order had not been passed down, but it could come at any moment. No one was certain who would take responsibility for the final attack command, but Acting FBI Director Patrick Gray was in personal contact with the field forces. He knew every situation had limits, and when they were reached, few if any courses other than violence were possible or advisa' le.

Jackson, who had his first order to crash thwarted, renewed his resolve. As they approached 1:00 P.M., he became more and more agitated. This time it was for real.

The management at Oak Ridge, alerted through their security team about midnight, set up a command post in the AEC building inside the installation. Realizing the threat was serious, a group of specialists from the major contract operator, Union Carbide Corporation, Nuclear Division, was alerted and called to duty. Shortly after 7:00 A.M. they ordered a shutdown of all plant operations which could be curtailed without "massive effect." Three research reactors in Oak Ridge were closed and many of the operations deemed hazardous in the Y-12 weapons plant were stopped. The gaseous diffusion plant, however, continued in operation.

Communication links were installed between the command post and local, state, and national law enforcement authorities. Contact was also established with the Knoxville Airport control tower, which was capable of receiving the hijacked aircraft's radio transmissions.

The public information section of the installation's staff drafted a short basic statement to answer the incoming queries. It consisted of a single question and answer.

Question: "What would happen if a commercial airliner were to crash into one of your plants?"
Answer: "If an airplane should crash into an Oak Ridge AEC facility it would have similar effects compared to the same kind of crash into any large industrial complex. Such a crash could *not* cause a nuclear explosion. In

the case of certain facilities, a crash of this nature could be accompanied by a localized release of radioactive materials, but this would not be a hazard to the public outside the plant affected."

To appreciate the full impact of this statement, some idea of the complexity of the installation is necessary. Oak Ridge is a 95-square-mile site near Knoxville, Tennessee, which contains a community of about 30,000 and three plants. The Oak Ridge Gaseous Diffusion Plant is the biggest and covers almost a square mile of land. In addition, this one installation has five large processing centers, each between 20 to 40 acres in size, plus another 70 auxiliary buildings. The Y-12 plant is located on a separate 500-acre tract and serves primarily as a nuclear weapons manufacturing, test, and design center. Some Oak Ridge National Laboratory activities are also carried out on this site, but the Laboratory has its own 500-acre development nearby, which contains several small research-size reactors, chemical processing plants, particle accelerators, waste disposal test operations, and other nuclear-related activities.

The massive size of the area provided the hijackers with an ample selection of targets if they could identify the functions of the various buildings from the air. But it was felt the very magnitude of the installation, unless highly unusual atmospheric conditions might accidently prevail at the time of the crash, would confine the damage caused by the radioactive plume to the Oak Ridge area. This basic concept was a part of the early planning for the site and was the justification for the statement "but this would not be a hazard to the public outside the plant affected."

All during the threat, AEC management maintained a calm front in their reaction to the crisis. They did not believe the hijackers would actually manage to drop directly into a reactor, but could tell, by monitoring the conversations between the three men and the tower, that the situation was "deadly serious."

Evacuation of the various plants, while not an everyday occurrence, took place routinely. (Only a month after this incident, the 5,800 people working in the Y-12 weapons facility were called out again. This time, it was a bomb scare. The ensuing search proved fruitless.)

The constant drone of the aircraft overhead, as it slowly

circled, served to agitate the residents of the area. The AEC requested additional telephone operators as a part of their well-developed emergency plan and over ten thousand calls were handled during the five to six hours the situation lasted. The local radio station began to broadcast the official release to ease the populace's concern, and had the good sense to refuse the services of an unidentified scientist who offered to come down to the station for the purpose of "evaluating" the threat to the community.

The prompt action by the AEC staff rapidly mobilized the installation. After a flurry of excitement and action, during which people were required to perform their tasks quickly and well, they were faced with the hardest chore of all. What could be done on their part had been. The only thing left was to wait it out. And so the tension mounted. The conversations between the hijackers and the authorities became more and more heated. As the final minutes of the extended deadline drew nearer, the three men on board became more determined. They realized they had made the ultimate threat and if they failed to follow through, the effectiveness of further coercion would be badly weakened.

On board the aircraft, the situation was desperate. Jackson, Moore, and Cale stalked the aisles. Passengers were threatened, but no one was actually assaulted. The nervousness of the three was clear for all to see as the deadline approached. Later, one passenger, who had remained calm throughout the ordeal said, "There is no question about it. They were going to do it."

The skyjackers maintained a constant battery of questions over the radio concerning the ransom. They pulled the pins on their hand grenades and showed them to the now thoroughly frightened passengers and crew, promising immediate annihilation. Jackson, gun in hand, moved forward into the cockpit as the minutes ran out and pointed his pistol at Captain Haas. The time had come for the final act.

The staff in control of Oak Ridge was in a high state of apprehension. As the 1:00 P.M. deadline drew near and time ticked away, more experts had been moved in to give professional assessments of the potential damage the huge aircraft might cause. They agreed with the issued statement but could not achieve unanimity on the amount of radioactive gas which might eventually find its way into the atmosphere. The meteorological staff was able to provide reports on the potential direction and velocity of drift from the various places on the site where the final thrust could come, but had trouble in estimating the total effect of the winds aloft on the outcome.

Jackson, in the closing minutes of time allotted before he would be called upon to make good his threat, acted with great deliberation. Thwarted once, he had, during the past hour, bolstered up his nerve. He was determined to go through with his part of the drama. Haas, still working against all hope to save both his people and his plane, waited while controlling the DC-9 in a precise circle some 20,000 feet off the ground.

The climactic moment had been reached. No time was left. Jackson was on the verge of giving the crash order when, without warning, the radio broke into static and a voice came through. The three skyjackers had won. Their wishes were in the process of being granted and a special Southern Airways plane was enroute to Chattanooga, Tennessee.

Jackson and his friends were skeptical. They had been lied to once and were not going to be taken again. Additional radio messages were exchanged. The three kept asking how much money was on board, and if their other desires had been met.

Finally, Haas asked the question. "We don't know how much money you've got on board, and the people want to know."

"Are you checking the amount?" Haas was forced to ask again.

"We have the amount requested on board. Repeat, the amount requested."

At this point, jubilation broke out among the three. The $10 million was on its way to them.

The truth, however, was the Southern Airways plane was carrying only $2 million in cash. They had the helmets, bullet-proof vests, and other items agreed upon, but they were eight million short of the needed ransom. It had been no small feat to collect the two million over the weekend, but the airline had managed. At the advice of authorities, who felt the ship would rapidly depart for Cuba as soon as the money had been transferred and the three on board would have little time to count such an immense amount of cash in small bills, the two million was considered to be sufficient.

As Captain Haas started a let-down toward Chattanooga,

71

southeast of the Oak Ridge complex, he made a valiant plea for the release of the other hostages. He personally offered to stay aboard and fly the men to any destination they desired but argued for immediate freedom of his passengers while they were on the ground. An agreement was reached, but as they approached the airport, the three were spooked by seeing thousands of people gather on hilltops and nearby shopping-center parking lots to view the spectacle.

"Pull down the shades," Moore was shrieking as he ran down the narrow aisle. The plane was on final approach to landing. "Get your heads between your legs. No talking." He was strident in his demands.

Looking out the windows, the three, in their imaginations, saw a potential law enforcement agent behind every man, woman, and child. Their earlier elation changed to agitated fear. They realized their safest course was to be airborne as soon as possible. Unloading the passengers would take time and expose them to the view of crowds and agents. So they decided no one would deplane and the final transfer and refueling would take place quickly.

According to prearranged plan, a fuel truck and a smaller vehicle moved with deliberate slowness to where the plane had stopped. The area around the end of the runway, far from the terminal building, had been selected because of its isolation.

Jackson, gun and grenade in hand again, had taken station in the DC-9 cockpit, where he could observe, watch Haas, and hear all incoming signals over the radio.

In the terminal, there was mass confusion. One eyewitness said the movement of the agents, coats flapping open showing their pistols as they ran from place to place, was constant. Passengers awaiting flights were restricted to certain areas, and great efforts were made to place lookouts at every available point.

Action in Oak Ridge, while calmed by the landing, remained intense. There was no real guarantee the hijackers, disappointed by not getting the $10 million, would not resume their deadly circling when they took off again.

Hundreds of people in the Chattanooga terminal, aware of the drama on the concrete runways outside the air-conditioned building, pressed for a look at the scene below.

Fred Vogt, stripped of his clothing and wearing only undershorts, climbed into the cab of the yellow refueling vehicle. He started his engine and, followed by the smaller truck, commenced the long, slow, crawling drive out to the waiting DC-9.

In the control cabin, Jackson and one of the other men waited with growing annoyance. Their plan called for everything to be passed up to the open copilot's window, thus prohibiting anyone from actually coming aboard.

As the fuel truck pulled alongside the wing to position itself for filling the first tank, the smaller vehicle came to a stop. Working from an elevated platform used to load food-service and maintenance personnel into a parked aircraft, Vogt handed through bundle after bundle of currency and the requested materials. On the wing, the man working the fuel hose was nervous and slipped, allowing the steel nozzle to bang on the main spar. Deciding the refueler was an agent in disguise, Jackson cocked his chrome-plated .38 and took aim.

"I'm gonna kill him." He spoke in an excited voice, and Haas, still in his captain's seat, answered, "He's just nervous in this situation." Then, taking a chance, he leaned forward into the field of fire and called out the window to the man, "Take it easy. You won't get hurt. Just get the gas in."

When the last of the bundles came aboard, the three were ready to leave and watched the lone fuel truck operator carefully. Was he delaying them? It seemed so, but there was no hurrying the pumps which delivered the JP mixture to the waiting tanks.

Then, after an eternity, it was done. As soon as the nozzle was removed, Jackson gave his order. "Let's go. Take her to Havana. And tell them we want Fidel Castro waiting at the airport."

With the ground wires still connected, the engines burst into a roar of life, and the plane was moving again. The copilot, Johnson, duly transmitted the last demands over his radio headset.

The takeoff was given immediate clearance and a corridor was cleared for the DC-9 by the most direct route to Cuba. The plane began its long takeoff roll.

Inside the terminal, the agents on duty realized their prey was about to escape into the air again and a shout rang out. Emergency vehicles filled with armed men were quickly started, and tires howling, went screaming after the ship. Several officers who had tried to get close enough for a sniper shot were left

73

stranded as the plane, in a shimmering wave of heat and kerosene-smelling black smoke, moved farther and faster away from them.

A brief chase ensued. Agents in cars and an ambulance, which had been commandeered on the off chance there might be an opportunity to board and remove the passenger undergoing a heart attack, sped through the bright afternoon. But it was too late. They watched the jetliner gather speed, rotate into a takeoff attitude, then lift into the air.

The hijackers had slipped away again—this time to a potential sanctuary.

The story doesn't end here. The plane, with the same complement of passengers and crew, flew to Havana, where Castro was waiting in a terminal building office. He refused to see them, however, leaving all dealings to a middleman to whom he passed instructions.

On the way south, Jackson, Moore, and Cale were jubilant. Their gamble had paid off. Wearing the helmets and flak vests, they paraded in the cabin before the watching passengers. Threats were still made, but this time not aimed at Oak Ridge. Now they were declaring they would don parachutes, set off a couple of hand grenades in the confined space to demobilize the flying ability of the aircraft, then bail out into the sea. But these boasts turned out to be idle.

The money, about two million dollars, was spread out "all over the rear seats," and the three, when not shouting or bragging, would run their hands through it with glee. It was more cash in one spot than any of them had expected to see.

Time and again they offered and gave money to the passengers. One man, who received $250, had no idea what to do with it and stuffed it into his shirt pockets.

The skyjackers started a party, washing down the amphetamine pills, which had come aboard as stimulants for the two pilots, with small bottles of liquor from the unlocked cabinets. This combination was later to put them through a bad series of elations and depressions.

One of their prize possessions was a "grant" from the President of the United States, stamped with a variety of impressive seals, and totally worthless. Moore, waving it in the air, said over

74

and over again, "Ain't this great? We're millionaires. The money's ours."

The various sacks and bags seemed to intoxicate the man, and on a wave of emotion, drugs, and alcohol, he turned benevolent. "Folks, we ain't got nothin' against you. Some of you businessmen have missed a day's work. You all have things you want to buy, and you've got payments on your houses. We're gonna share the wealth." With that, he began another round of passing out dollars. The passengers were afraid to refuse.

Moving into the cockpit, he began piling bills around Jackson and Cale. By the time he had reached more than $200,000, the money was so deep Captain Haas had to request he stop, because it was interfering with the controls.

The hijinks went on for the entire trip. During the time in the air, the men would go from high spirits to deep depression.

The radio remained in constant use relaying progress notes on their whereabouts and conversation among the hijackers, the pilot, and the authorities on the ground. The three remained firmly in control.

As they neared Cuba and Haas began to make his clearing turns for José Martí Airport, Moore became more cautious. Once more he demanded all shades be drawn and the passengers bend over, heads between their legs.

The reception of the skyjackers in the Cuban capitol was less than cordial and far less than the men had expected. Whatever deal they had intended to discuss was forestalled by the hostile actions of the Cuban troops who surrounded the plane.

Moore and Jackson, still carrying their pistols and hand grenades, climbed down out of the aircraft by throwing an escape rope out the cockpit window. A great deal of shouting ensued. Apparently none of the Cubans spoke English and neither of the hijackers could communicate in Spanish. Instead, in an effort to use volume as a substitute for understanding, they yelled at one another. The Cuban guards and officials began to close in on the two men, who scared them back by waving their guns and showing the grenades.

It was a hard climb back up into the aircraft, but the two made it. Then in a fit of rage at not having been welcomed to the Communist island as heroes for having hijacked a capitalist airplane, Jackson stuck his head out the window and gave vent to his bitterness.

"Get me Fidel. And get me a fuel truck, or we're gonna start throwin' dead people out of this airplane."

Moore, meanwhile, was talking to the passengers. "Those people wanted to arrest us," he said indignantly. "Can you believe that? Why, they're nothin' but a bunch of Spanish-speakin' George Wallaces."

A single passenger on board could speak Spanish, and he was brought to the cockpit. In an emotional plea, he spoke with the tower asking for fuel and stating the seriousness of the situation. "For the love of God, comply with their demands," the man said passionately. "This situation has been going on over twenty-two hours and the hijackers are desperate."

The Castro forces, in answer to this request, furnished the jet with a minimum of fuel, and the plane was quickly airborne again. With only partially filled tanks, it could not remain so for very long.

A debate commenced, and after several radio transmissions, a landing at the U.S. Naval Air Station at Key West, Florida, was agreed upon. This was one of the closest places where the type of JP fuel compatible with the DC-9's engines could be found. The time on the ground was brief, and the completely refueled plane was once more in the air.

Jackson, holding wads of money in his hands, instructed the pilot to take the plane to Switzerland. That nation's record of neutrality seemed to make it the only possible haven.

Captain Haas, however, vetoed the idea by explaining the DC-9 did not have the necessary range for a transatlantic hop, and that their engines were running low on a special oil. Still not convinced, Jackson ordered the plane to land in Orlando, Florida, to "get what you need."

On the ground, a major gamble took place. The FBI agents, under orders from Acting Director Patrick Gray relayed through his agent-in-charge at McCoy Air Force Base in Orlando, were instructed to stop the plane from taking off again at any cost. The decision was made to shoot out the tires and then, when the ship came to a halt, board, using techniques a special team had been practicing all day long.

Although J. Edgar Hoover had a long-term understanding

with the Air Line Pilots Association about action without the pilot's prior knowledge or consent, Acting Director Gray decided not to contact Haas and use the code to tell him of his plans. He felt this decision was justified because he did not consider the pilot and copilot "free to exercise their best judgement."

In any event, the FBI and other authorities wanted the plane grounded before another Oak Ridge incident could occur. There was a good possibility the trio, having once been successful with that ploy, might try again—and this time might actually crash.

While the refueling was going on, fifteen agents crept through the darkness and gained sanctuary under the fuselage at the rear. As the fuel truck roared away, they opened fire on the tires of the DC-9.

Dim pops were heard inside the heavily soundproofed cabin, and the aircraft lurched off to its left as the tires on that main gear were blasted away. Although possibly hit, the nose wheel and right main gear tires were not flattened.

Moore and Jackson, standing at the open cockpit door, were at first puzzled by the noise and the sudden tilting of the plane; then, realizing what had happened, they went wild. They fired at the rapidly departing fuel truck, then shot again through the cockpit windows, past Haas's and copilot Billy Johnson's heads. Still insanely angry, they started shooting through the galley floor at the men they knew were lurking under the aircraft. This is when the only incident of intended injury to passenger or crew occurred.

In a complete rage, Jackson pointed his pistol at Johnson, who had been handling radio communications with the tower, and started screaming. "We're gonna start shootin' people and throw 'em out the window. And you're first, Harold." He had been mispronouncing Billy Halroyd Johnson's name for hours. "You did it," he screamed again. "You told 'em to do it, and we're gonna kill you."

Johnson, realizing his life was in danger, resisted as the two hijackers dragged him out of his seat and into the main cabin. He flung himself to the floor between the second and third rows and Jackson fired. The first shot exploded in the confines of the passenger space, and the bullet blasted through one of the seats into Johnson's arm. The man screamed in pain, and as Jackson tried to fire another shot, he rolled partially into view. Jackson's chrome-plated pistol failed to go off the second time he pulled the trigger, and before he could try again, Lou Moore stopped him. "That's enough, man. Don't kill him. We may need him. Let's get out of here."

Mollified, Jackson motioned with the barrel of the gun. Johnson, pale-faced, with bright crimson blood pouring from his wound and staining his once white shirt, climbed slowly to his feet. Jackson pushed him roughly into the cockpit, and began hollering again to "Take off."

Haas argued the impossibility of such a maneuver with the tires blown away, but was again faced with a threat. "Get into the air, or we'll start killing people. Harold will be first."

Haas slammed on the power, and the agents under the aircraft were blown away by the force of the jet wash. One man rolled and tumbled half the length of a football field before he could stop himself. When he stood up, his clothes had been torn away.

The plane, lurching drunkenly, rumbled down the runway until, with a great reeling bounce, it was airborne. The treads from the flailing tires were sucked into the engines as they flew off the wheels, and the rims, trailing showers of red-orange sparks, dug parallel lines in the hard concrete surface.

The odyssey, which had now gone on for more than 24 hours, was still far from ending. The plane, with no rubber on the left main gear and engines badly strained by their ingestion of the flying parts of the shot-away tires, had only one landing left. Haas, who had been at the controls the whole time, was now without the assistance of his copilot, who had a shattered bone in his arm.

The hijackers, still in a temper, tried to talk to the President but refused to converse with Secretary Volpe of the Department of Transportation. Some thought was given to crashing into the Key Biscayne home of Nixon, but the money and the possibility of better treatment in Cuba drove them to their final selection of a field. They ordered Haas to return to Havana.

Using the passenger who spoke Spanish to request a foam-lined landing surface, they were amazed to learn there was not enough of the fire-preventing material to cover an entire runway they would have to be met by a foam-fire truck when they stopped their forward motion on the ground.

Captain Haas, after orbiting the José Martí field long enough to burn away almost all the remaining fuel load, brought the plane in safely.

The three hijackers, Moore, Jackson, and Cale, stuffed plastic flight bags with money. Each waited by an exit to make a fast getaway as soon as the plane stopped.

The stress imposed on the airframe during the landing was enormous. Flying with great concentration, Haas brought the tireless left gear in first, throwing up a cascade of sparks from the cement. Then, holding direction perfectly, he lost more speed and eased the ship down onto its good gear. Metal screamed as it tore and the plane shrieked with the twisting fuselage. Then, after a rush of noise, there was silence. They were down. The three hijackers burst from the plane and ran, trying to hide in the tall weeds near the runway. Three minutes later, they were prisoners of the Cuban Government.

The passengers were personally greeted by Castro, who was sincerely glad for their safety. They were treated well, and after a good night's rest and medical care for those in need, were allowed to return to their homes.

The hijackers were another matter. No one is certain what deal, if any, was worked out. For several years, the three have not been officially reported to be in the United States. Informed opinion indicates they are being well treated in a Cuban minimum-security prison.

The ordeal, which holds some kind of record for skyjacking, lasted more than 30 hours. The plane flew over 4,000 miles, landed nine times, crossed and recrossed the borders of three countries, and at one time or another had over 1,000 people involved in some phase of either flight direction or apprehension.

The threat to the Oak Ridge complex caused great anxiety for a prolonged period, and almost became a full-fledged nuclear incident.

All in all, it was some ride for the thirty-one people aboard Southern Airways flight 49. Everyone involved will remember it for a long time—especially the people at Oak Ridge. Because they came very close to a moment of destiny.

■ One of the main causes of the proliferation of nuclear installations on a worldwide basis has been certain policies of the United States Government. A step-by-step tracing of U.S. nuclear political history since the early 1950s will more than support this view.

"Atoms for Peace" was established in 1954. Under this program, any nation more or less favorably inclined toward the United States could have a nuclear reactor erected in a location of its choosing. The Atoms-for-Peace plan was viewed as a powerful tool by which the U.S. could control the spread of nuclear weaponry and at the same time support the economies of world nations by making a positive contribution to their needs for energy. It was eventually to have effects producing opposite results.

No one can dispute the need for Atoms for Peace. Nuclear reactors were and are expensive. One being erected in Arizona will reportedly cost more than the total assessed value of the city of Phoenix. If the underprivileged, technically less-experienced nations were ever to share in the bonanza of atomic energy, some more highly developed nation—and that meant the U.S., the only real atomic power of the time—had to come in and help.

Few political hooks were attached to the grant of a nuclear reactor. Few were needed. The Atomic Energy Commission regulated more than 90 percent of all supplies of enriched uranium utilized by the non-Communist countries of the world. And this control of the one vital element gave us practical dominance over all the U.S.-donated installations. Inspections were on our terms, as was the enforcement of regulations concerning the disposal of waste materials like plutonium. In short, that which we had granted we could also take away, by reducing or cutting off the flow of uranium. The deal was too good to last.

In the early 1970s, the Nixon Administration began a series

of steps which would eventually result in massive proliferation of atomic reactors outside the control of the U.S. government.

Under a loose program known as "privatization," private vendors were encouraged to become suppliers of nuclear materials, and the federal powers pulled back to reduce the importance of the Atomic Energy Commission.

This new move resulted in the 1974 shutdown of the AEC, which became, in part, the Nuclear Regulatory Commission, and turned the supply function for all previously regulated atomic raw material over to large, publicly owned corporations.

Sensing a period of potential price instability and fearing cost increases when the uranium industry was freed of government control and subsidy, a last-minute panic among nuclear power plant operators doubled orders for enriched uranium. This tied up all of the then-available AEC supplies. An artificial shortage ensued, and the price of most atomic elements rose with sudden sharpness in the world markets.

Coinciding with this seemingly accidental problem came the unforeseen activism of environmentalists and others moving in unison against the construction of new nuclear facilities. A ground swell of public response occurred almost as if in reply to a single signal. Overnight, groups were vocally and visually active in numerous countries, causing both politicians and private businessmen to move with great caution. The outcome was a dramatic slowdown in the growth of new nuclear installations, and a reappraisal of the long-term prospects for atomic power. At the time of this writing, industry estimates show no plans for the construction of more reactors after 1985 or '86.

Several major companies which had invested millions in the development of process plants to produce nuclear fuel elements—in direct response to the stimulation of an apparent shortage—were left in the unenviable position of having made large capital outlays to produce materials for which there was a declining market.

General Electric and Westinghouse, according to industry reports, had laid out over $500 million each in pursuit of this phantom market. Others, in joint projects, had lesser but still significant dollar investments. Combustion Engineering and Babcock & Wilcox are reported to have had combined costs of over $150 million.

The hardest thing for a business to do is turn its back on a major capital outlay for a production facility. Usually, there are possible conversions which, providing funds are available, can be made. Almost every industrial and chemical process lends itself to this kind of flexibility. In fact, initial design usually takes a duality of purpose into consideration so a large plant is seldom constructed without serious study given to alternative uses for the complex.

A nuclear facility, however, is different. The federal and scientific safeguards, along with the ever-present problems stemming from radioactive contamination, render the design single purpose.

In other words, if you invest $200 million to construct a facility for the production of enriched uranium, what you end up with is precisely that, and nothing else. You cannot switch it over to manufacture alcohols or ammonia.

Faced with a hard decision, the business firms which had joined in the Nixon-supported privatization began to look for ways to generate some return on their now-doubtful investments. Shortly before the demise of the AEC, two U.S.-based nuclear brokerages were opened, to bring buyer and seller together in an open market.

The World Nuclear Fuel Market (WNFM) came first, and the Separative Work Unit Corporation (SWUCO) followed. These two firms arrange sales across international borders. SWUCO, with headquarters in Maryland, operates as a listing service. A seller lists his commodity, and a buyer his needs. With this information available, deals are consummated.

WNFM acts more like a consortium. In 1976, there were 79 nuclear-oriented business organizations, ranging from construction contractors to power utilities, as members. Thirty-eight were U.S. companies and the forty-one remaining were from other countries. A recent report indicated SWUCO has assisted in the dissemination of enough plutonium to construct eight or nine atomic weapons. One shipment was said to have been sent from a U.S. company to an overseas buyer.

The existence of these two firms provides interested U.S. companies with world markets for their nuclear produce. Some of the nations buying considerable quantities of atomic materials would have been held in check by the old AEC. The new NRC has no responsibility in this area.

In short, there is today a sort of gray market in which U.S.

companies are making sales to nations with questionable motives or with less-than-strong allegiance to the United States and the non-Communist world.

The need to participate in this rather shady arena stems from the series of events previously recounted. And it is a major source of massive proliferation.

Some open shopping for atomic weapons has taken place. President Quaddafi of Libya has reportedly been trying to buy a bomb at any price. To date, as far as is known, he has been unsuccessful. But since money is no object, there is little question about his prospects for success if he perseveres.

Another report, concerning a confidential memo from the Energy Research and Development Administration (ERDA) to a congressional subcommittee, states that at least twenty-two sales of nuclear materials and artifacts have been made to foreign countries by U.S. companies since the onset of privatization. These do not include the secret aid reportedly rendered by the CIA to selected governments.

The competitive pressures on American firms with high investment in the nuclear field have been intensified by the formation of several consortiums of foreign governments and corporations.

A single month's reports show Brazil entering into an agreement with West Germany, and through the West Germans, with Holland, for reactors and plutonium reprocessing. Pakistan signed a contract with France in 1976 for a reactor, postponed the agreement due to "political unrest," and after a military coup in June resulting in the removal of Prime Minister Zulfikar Ali Bhutto, the French attempted to stall the deal. When the Pakistanis indicated they would go to "other sources," believed to be a reference to their close ally China, the French came through. France is also pursuing the sale of breeder-type reactors in the United States. One thing we have achieved through our policy of vacillation is the end of our technological supremacy. Which in turn makes already bad matters worse by giving everyone who wants to make secret or hard-to-trace nuclear purchases the added anonymity stemming from the competitive market, buyer-seller relationship.

The worldwide control of atomic materials has slipped out of our hands. And while this might have happened anyway, the removal of U.S. governmental regulations through the termination of the powers of the AEC clearly hastened the situation.

In the late 1940s only one country had the bomb and the ability to harness the atom for peaceful uses. In the 1950s, two more nations joined the team of atomic sisters. Then, in the 1960s, several gained the necessary technology. Now it is a pretty unruly and diverse family. Nine nations, some of them real surprises, have or will very shortly have some form of atomic explosive device. In addition to the U.S., the Soviet Union, the United Kingdom, France, China, India, Israel, South Africa, and Taiwan are members of or applicants to this not-so-exclusive club. Six more nations (Australia, Belgium, Canada, Italy, Japan, and West Germany) have a nuclear weapons capability but no compelling need to develop a bomb. But at least twenty nations are capable of making a nuclear weapon within a few years without help, or in even less time if aided. And more are applying daily.

Sooner or later some not-too-disciplined, over-egoed demagogue heading a one-man dictatorship is going to have the capacity to set off a nuclear explosion. Unhindered by problems of morality, and driven by the necessity of preserving his petty power, he is going to unleash a bomb. A lot of people are going to meet an untimely death. The radiation content in milk in Minnesota is going to go up. Hens' eggs in the Ural mountains are going to cause a slight but perceptible increase in clicks on a scintillation counter. Because we have only one earth. And even a "clean" bomb of reasonably small proportions makes a big enough mess to contaminate and pollute the atmosphere for a measurable period of time.

Regrettably, a lot of the problem can be traced to our doorstep. It is not yet too late to do something about it, because all of this is a part of the legal international movement of nuclear products. But it soon will be. This is the gray market—not the black market, which may also exist.

In 1976, the GAO reported government-owned atomic installations could not account for slightly more than 11,000 pounds of plutonium and enriched uranium. This is called missing or unaccounted for materials (MUF). Worse, a yet-to-be published GAO report is said to have come up between 15,000 and 20,000 pounds short in an audit of privately owned manufacturing plants.

In 1977, however, the government announced that 8,000 pounds were missing from nuclear facilities in the U.S. Their figure did not include weapons facilities, for national security reasons. John D. Dingell (D-Mich.) said according to an independent GAO audit, the government substantially underestimated losses.

That is at least 8,000 pounds of special nuclear materials which have only limited use. They can fuel reactors or build weapons. The four tons had to go somewhere. And since risk is required in the removal of the missing uranium and plutonium, the price of this MUFfed material on the world black market could amount to millions of dollars.

A portion of the unaccounted-for critical elements may have passed into the hands of the CIA for dispersal to selected overseas allies based on secret governmental policy. Another small amount was anticipated to be missing, as it is lost in the processing system. Inventory control procedures, record-keeping errors, and other factors may account for still more of the total. But it appears some is really gone, apparently without a trace.

It is difficult to determine exactly what has happened to this missing measure. While rumors about a black market arise from time to time, no concrete evidence exists to support this highly plausible speculation. Many unofficial statements tend to a theory in which the major producing firms have held out certain amounts awaiting price increases in the gray or legitimate markets. A third possibility is contained within the congressional testimony of Dr. Arthur Tamplin, a man admittedly opposed to the use of nuclear power. "It's certain," said Dr. Tamplin, "that the MUF accounting methods are not adequate to prevent the theft of strategic quantities of this material [plutonium]. They are not necessarily even adequate to detect the loss of that amount of material; that's why they have what's called the 'limited error of MUF!'" Outright theft of the radioactive substance appears to be possible. The NRC, however, denies any great losses over and above those caused by accounting or processing. And they defend their position with logically presented evidence.

The question, however, lingers. It has even been presented as the reason behind the death of an obscure technician. Which brings us to the next example of modern nuclear drama. It's the strange and some say still unsolved story of Ms. Karen Silkwood.

？

By 7:30 on a cold November evening, the last vestiges of light had fled from the dark sky and the first few stars, their brilliance not diluted by city lights, shone with sharp intensity. The narrow farm road cut a black strip across the silent, gently rolling Oklahoma countryside.

A lone girl wearing a Mickey Mouse wristwatch sat behind the wheel of a white '73 Honda Hatchback, boring through the night to keep an appointment set days before with a union leader and a reporter from *The New York Times*. Beside her in the other bucket seat, a large brown manila envelope moved slightly as she corrected the car's direction. It contained the final, much-needed evidence required to back up the statements she was soon to make about large amounts of missing plutonium.

The headlights cut a white corridor in the blackness along Highway 74, revealing the compact shoulders and the narrowness of the thirty-four-foot-wide asphalt strip.

At 55 mph, the wind noise inside the small car blocked out all other external sounds. She had been driving about 10 minutes and was 7.3 miles south of Crescent, Oklahoma, a small farming community where she lived and had friends.

The driver was relatively new to the area. She came from over five hundred miles to the south; from a bleaker place of brown, flat, featureless terrain which smelled of refined oil and petrochemical production. But she lived here now, sharing an apartment with a roommate and working in the Kerr-McGee Cimarron plutonium production facility.

The road, following the contour of the land, turned slightly downhill, and without thinking, she adjusted her speed.

Suddenly, from nowhere, the blinding flash of another car's headlights reflected off her rearview mirror, causing her to squint and turn her head away from the annoying illumination. She allowed her Honda to move closer to the shoulder, giving the overtaking vehicle the maximum possible room to pass.

Unable to ignore the intrusion of the intense light, she shifted in her seat. The other car drew closer.

She gave a quick glance in her mirror. The second vehicle was directly on her tail. Vexed by the closeness of the other auto, she came down with her foot and increased speed. The second car followed.

Then, without warning, she felt a sickening lurch as the driver of the second automobile deliberately nudged her left rear bumper.

Once was enough. The wheel, free of the ground for an instant, spun madly, then came back into contact with the oily pavement. The nose of the small car slid to the right. She managed to correct the skid but applied too much countersteering and the car, crossing over the broken centerline of the narrow roadway, was almost out of control as it reached the verge of the hard-packed shoulder. The normal reflex action was to brake, but it was too late. Ahead, the gray-white square of a concrete wing abutment for an under-the-road drainage pipe loomed directly in her path.

The noise inside was insane. Wind and shrieking tires, then a single multisound crash as the Honda impacted with the immobile barrier. Rending metal and the shattering tinkle of breaking glass blended into a final cacophony. Then, resonating silence, broken only by the noise of the car which had caused the wreck as it slowed to a stop some yards further ahead. Gravel grated under the rear tires as the driver threw it into reverse and gunned the engine, backing rapidly to a position across the road from the now-shattered Honda.

A man emerged from the vehicle. Running lightly, he crossed to the culvert. Working by the light provided from the destroyed car's headlights, he fished inside the passenger space and retrieved the manila envelope. Rummaging through it, he tossed papers into the air until he found the ones he sought. Clutching the envelope to his chest, he ran back across the roadway and jumped into the waiting car.

The driver, hesitating only long enough to hear the slam of the door, came down hard on the accelerator and they zoomed off into the night. In seconds the rapidly moving vehicle had crested the low hill a few hundred yards further down the two-lane blacktop and was gone. The final booming note of the exhaust was absorbed into the darkness, and it was still again.

Nothing stirred. Less than two minutes had elapsed from the point of the first impact between the cars.

The broken, bleeding girl at the wheel was dead. The documents she had so valiantly collected and tried to deliver, gone.

This is the version of the happenings on that Wednesday night, November 13, 1974, which a lot of people accept as gospel. And gospel truth it may be. We may never know for certain.

But there is another side to this story. Another reconstruction of the events—equally plausible and elegantly more simple. It goes like this.

A very distraught young woman, Karen Silkwood, completed a telephone call shortly after 6:00 P.M. to remind her boyfriend, Drew Stephens, to pick up Steve Wodka, a union official from Washington, D.C., and David Burnham, a reporter for *The New York Times,* to deliver them for her scheduled meeting at 8:00 that night at the Holiday Inn-Northwest in Oklahoma City.

Satisfied that all was in order, she left the Hub Café, and driving slowly, made her way to Highway 74.

The girl with the Mickey Mouse watch was in a highly emotional state. On the previous two days, Monday and Tuesday, November 11 and 12, she, along with Stephens and her roommate, Sherri Ellis, twenty-two, a fellow worker at the Kerr-McGee facility in Cimarron, had undergone stringent medical examinations at the Los Alamos Scientific Laboratory in New Mexico to determine how badly they were contaminated by plutonium.

They returned home between 10:30 and 11:00 P.M. on Tuesday, and spent some time consuming a number of Bloody Marys made from tomato juice and 190-proof alcohol. The liquor and Methaqualone, a sedative prescribed by a doctor some weeks before, make a powerful combination. No certain testimony exists as to the time Karen went to bed, but according to her roommate, it was later than 2:00 A.M.

The day had been arduous enough. After a morning-long meeting which involved her in contract negotiations between Local 5-283 and the Kerr-McGee representatives, she had gone through a tiring session with investigators from the

Atomic Energy Commission who were trying to run down the source of her contamination. According to the testimony of FBI agent Bill Fisher, an AEC inspector was interviewed by one of the official investigators of the accident and he reported the girl appeared emotionally upset at times during their conference. On one occasion she had broken into tears.

A great many things must have been running through her mind as she turned her car onto the farm road and settled back for the 30-mile drive to Oklahoma City. A lingering fear of permanent injury from the plutonium, the stymied union negotiations, her desire for better working conditions, her restriction from radiation work because of the contamination, and a general lassitude from the combination of drugs, lack of sleep, and the aftereffects of alcohol.

Little traffic passed on the dark, silent road and the headlights bore hypnotically into the soft, almost palpable blackness. What started as an agitated, excited mind switching from subject to subject became, in minutes, a brain gone random, absorbing the darkness and the tempo of the highway.

The first gnawing of drowsiness bit painfully at the edges of her consciousness and she shifted her position in the narrow bucket seat.

She settled back again and involuntarily allowed the return of fixed attention, then blankness before her eyes.

The small white car, following the steering input, veered slowly to the left, over the centerline, edging closer and closer to the shoulder.

Something—a natural awareness causing a return of her consciousness from the realm of half-sleep, or the change in the sound of the car tires as they moved onto the very edge of the road—brought her around once more. But this time she was confronted with a moving emergency. Still not fully alert, and working with the hyper-reflexes of an adrenalin-charged system, she started to swing the wheel. Closing at a rate of almost 60 feet a second, the white Honda shot toward the concrete abutment. What had started as a potentially serious accident had now become a deadly one. The impact threw the passenger violently forward.

Minutes later a trucker, following another car, came onto the scene and spotted the crumpled wreck. Stopping to investigate, he found the girl's battered remains still inside the overturned vehicle.

Papers were scattered about, and among them there might have been a brown manila envelope. But no one is certain. There were no documents indicating massive losses of plutonium in the Cimarron plant found among her effects.

Within hours of her death, Steve Wodka notified the AEC of his union's suspicions that Karen might have died in the car crash as a result of foul play. Because the highway patrol investigating the accident felt it was merely a case of falling asleep at the wheel, and the union could find no governmental agency to investigate their allegations, Wodka initiated an investigation.

The accident investigator hired by the union saw the car in Oklahoma City, where Drew Stephens had had it towed, two days after the accident. He found damage to the left rear of the Honda which he considered sufficient evidence Karen's car had been hit from the rear and forced off the highway.

One other person who had seen the car shortly after the accident said there was no damage to the Honda in that area.

The death of Karen Silkwood, coming as it did at a time of union unrest, and after her mysterious contamination at a plant with known AEC violations, has created a cause célèbre for the antinuclear side. And by now, regardless of the truth, she is regarded by many as this country's first atomic martyr.

But the main significance of her death lies not in the mystery of the events surrounding her violent demise but in the information produced by the many lengthy inquiries into those circumstances.

Karen Silkwood, born in Longview, Texas, grew up in the small town of Nederland, located in the center of the Gulf Coast petrochemical production area. At school, she was a member of the Future Homemakers of America, a flutist in the band, and a member of the National Honor Society. She was interested in science, especially chemistry, and took a special course on radiation.

After a year at Lamar College in Beaumont, Texas, she met a pipeline worker while on vacation. They married and for the next six years lived a nomadic life as he moved from place to place across Texas and Oklahoma, following his trade.

The end of this relationship came after three children.

91

Divorced, Karen drifted for a short time, then moved north to Oklahoma. Custody of the children went to her ex-husband, who had remarried.

After a short stint as a clerk in a hospital, she became, on August 4, 1972, a laboratory analyst in the Kerr-McGee plutonium-producing plant in Cimarron, Oklahoma.

This facility was built in 1970 to fulfill a multimillion-dollar federal contract to supply 18,500 plutonium fuel pins for the Fast Flux Test Facility in Richland, Washington. Located next to another Kerr-McGee operation, an automated uranium-reprocessing plant, the Cimarron works made pencil-thin tubes from metal, welded the seams, then filled them with plutonium.

According to company officials, Karen was, in her early days of employment, an excellent worker. She displayed a cheerful attitude and showed enough interest in her job to have considered it as a career.

Then, after about three months, things began to change. Karen, through her friends, became progressively more and more involved in union activities. Gaining quick recognition for her spirit and willingness to speak out, she soon managed to become one of the leaders in carrying workers' problems to management.

Why she found satisfaction in this role is unimportant. One group maintains she started to see the real plight of those who worked with radioactive materials and reacted by redoubling her activities and commitment to the union. Others say she had found a gratifying outlet, peer respect and recognition by associating herself with the leadership of a dissident group.

Whatever her motives, she soon became well known. Comments about her ran from "emotionally troubled divorcée" to "a good kid."

Her devotion to the union cause was matched by her antagonism toward Kerr-McGee, and especially the plant in Cimarron.

The plutonium facility was a problem to Kerr-McGee in many ways, including labor relations and operation safety. The Kerr-McGee Cimarron plutonium plant was an atomic installation of a unique nature, and at times the management at the site had difficulties in their dealings with safety regulations.

An examination of the accident record of any major manufacturing facility will reveal events in which workers were injured and those in charge failed to take the immediate steps specified in the plant operating plan. These same omissions, when coupled to the added problems of dealing with nuclear materials, become more serious at an atomic installation, but are still commonplace: when a number of people are working in a relatively small space and they handle dozens of duties on a daily basis, they will become careless, or casual. This relaxation of concentration leads to accidents.

One protection against this is constant training. And according to Anthony Mazzocchi's testimony before a House of Representatives investigating committee, the Cimarron facility had no such ongoing program. In addition, several workers who had received little or no training for their jobs were working in crucial areas.

A chronological listing of incidents and accidents at any site the size of the Cimarron plutonium facility gives the feeling of massive and consistent disregard for employee safety. While most of the items by themselves were of no great significance, collectively they indicated management controls being exercised should be strengthened.

Story after story listing plant failures has appeared since Karen Silkwood's death. Some of these reveal management mistakes and others, worker errors. Still more would seem to be attributable simply to chance.

In October 1970, shortly after the official opening, two men were contaminated when a storage container was mistakenly left open for three days. Then, three months later, in January 1971, twenty-two more individuals were exposed to radiation when a defective piece of equipment failed and allowed plutonium oxide to leak into the atmosphere of the work area. No other significant nuclear incident occurred until April 1972, when two maintenance men working on a pump were splashed with plutonium particles in liquid suspension. Not realizing their contamination, they left the plant at noon for lunch in a nearby town and did not become aware of their condition until they returned to the facility hours later. Both underwent successful decontamination processing and their car was cleaned up. But Kerr-McGee staffers reportedly failed to check the restaurant where the two had eaten, and neglected to report the matter to the AEC as required by the operating regulations. The AEC learned of the incident a month later and the affair, a violation of the federal nuclear code, was processed through channels. Mild corrective steps were reported to have been taken.

Other events during this period included the appearance of tiny holes in the gloves workers wore while handling radioactive materials inside the gloveboxes, drum leakages in the tanks used to store plutonium, process system errors in which plutonium flowing in solution through the complex system of piping was wrongly routed to areas of the plant not designed to receive it, and equipment failures such as an exploding compressor which killed one man. All of these incidents are regrettable.

Union officials in testimony before the House investigating committee have indicated improved training, along with closer adherence to the operating rules, might have prevented some of the trouble. Not unnaturally, the accident rate at the plant contributed to poor relations between the local of the Oil, Chemical and Atomic Workers International Union (OCAW) and Kerr-McGee.

Oklahoma, as much of the Sunbelt, is not a strong union area when compared with other parts of the United States. Traditions of Old West independence coupled with labor practices stemming from the founding days of the oil and gas industry have made union and management relations troubled.

The OCAW, in their bargaining attempts with Kerr-McGee for a new contract at the plutonium plant, came to the conclusion a strike was their only alternative. So, in late 1972, their members walked off their jobs.

Kerr-McGee responded by bringing in other individuals to fill the vacated positions. According to the union, there was no sufficient training program for the replacement employees. While no significant number of incidents was reported during the time the facility operated with these less experienced workers, the strikebreaking move served to further irritate union members and widen the gulf between the local's officers and plant management.

This is an unhealthy climate in any manufacturing plant and lends itself to the creation of constant strain, which adds yet another distracting element to an unstable environment.

Karen Silkwood's death gave the union a means of forcing the AEC to look at working conditions at Kerr-McGee. The AEC, in specific answer to the allegations and rumors about safety, conducted an extensive on-site investigation. Workers and management alike were called upon to come forward and testify. Documents dealing with the operation of the Cimarron facility were stringently reviewed.

In the final analysis, 39 safety grievances were investigated; 20 were found to be true or at least to have some basis in fact. Plutonium had been stored improperly—in one instance in a desk drawer, instead of the locked vault provided for the purpose. Workers had labored without wearing respirators in areas not tested for contamination as required, or in places where leaks had occurred. Kerr-McGee operating management did fail to report a May 1974 incident in which a serious leak closed the plant. And written standards had not been fully followed in checking and inspecting respirators. In addition, the AEC team found improper worker training in several areas of the facility.

Granted, none of these problems should have existed. But they are far from being criminal acts. In fact, the AEC did not consider them worthy of censure, and other than including the incidents in a file on Kerr-McGee and the Cimarron plant, no punitive action was recommended or taken.

Antinuclear groups have developed an elaborate explanation for this lack of censure, based on the fact that Cimarron was the production site for the plutonium fuel rods to be used in the still-under-construction Richland, Washington, breeder reactor. According to several articles, the AEC was afraid to take strong measures against the Kerr-McGee breaches of regulations because it would have spelled bad publicity for the breeder program and cast doubt on the success of the project. Others claim inordinate political influence was the source of a cover-up.

Dear as this logic may be to the hearts of those seeking a national conspiracy of big business and government, another explanation suggests itself.

In the period 1973–74, the AEC investigators found 3,333 violations of regulations in various nuclear establishments in the United States. Only 8 of these were considered to be sufficiently serious to permit a penalty. The balance, while still considered to be noteworthy, as is any rule breakage in an atomic installation, were deemed to be of no real consequence and were entered into the records of the plants and stations so future inspectors would be able to re-check from time to time.

In short, a rule infraction sufficient to result in a penalty was rare. It is therefore not surprising no steps were taken against the Kerr-McGee operations at Cimarron.

Now it may be that penalties should be given out for less serious violations. The NRC, in fact, has pressed for this. But no evidence exists to suggest the AEC was too lenient in this particular case, based on their previous history.

Karen Silkwood began work in the plant in August 1972. By November the strike was called. Nine weeks later, of the original 130 union members, only 27 were left to return to their jobs. Karen was among them. In April 1974, she was elected to the union's governing committee. On September 29, 1974, she was in the Washington offices of the OCAW, along with two other Cimarron employees, to discuss "strategy" with union executives Anthony Mazzocchi and Steve Wodka.

Wodka said he and his boss Mazzocchi, after hearing stories of alleged quality-control violations, asked "Karen to go back to the plant to find out who was falsifying the records, who was ordering it and to document everything in specific detail."

With urging and backing by the union, it is easy to see how this request became an obsession. Karen later vowed to her boyfriend Stephens she was going to get proof that Kerr-McGee management was falsifying its records in order to avoid trouble with the AEC. According to a published quote, she said, "We're really gonna get those motherfuckers this time."

Much of the information on quality control she gathered was common knowledge among both workers and supervisors in the facility. A retouched photograph of the welding on one of the fuel rods manufactured at the plant, often mentioned to show she had found evidence of falsification of records, also came to the attention of the AEC. They did more than Ms. Silkwood by finding the man who had doctored the photo to make a questionable seam appear to be passable. In an interview, this employee said he "used a felt point pen to dub a small amount of ink" on spots he considered "caused by defects in the emulsion [of the film], static electricity, etc." Even in this case, the welds had been visually inspected prior to the taking of the photograph, on a device called a metallograph, and the technician was "correcting" photos to pass a superior's inspection. Statements of the individuals involved indicate they were very concerned about controlling weld quality as they knew the potential problems which could arise from a weak tube.

The combination of the climate at the plant and Ms. Silkwood's attitude explains why she was so ardent in her investigation and so adamant she would find something.

In virtually every report of this case, Karen Silkwood has been shown gathering information about safety irregularities in the Cimarron facility. In fact, some of the union documents, written after her death, reinforce this idea. The reality, however, according to the statements of almost everyone involved, including the union officials, is that she was engaged in the collection of data which would document discrepancies in quality control.

Ms. Silkwood was known to have gone into private company files and copied documents she thought were of vital importance to the union's case. Those same files were also gone through by the AEC and were the basis for their validation of some of the previously mentioned regulations breaches.

The stress on Karen Silkwood during this period of her life was extreme. Yet, outwardly, she appeared to be coping with the situation.

She shared a small apartment, had a number of friends, partied a little, and was considered by many to be a "good listener."

But she had two significant problems. The first was her use of drugs. In addition to marijuana, common enough in our society in her age group, she also took regular doses of Methaqualone, purchased on a prescription issued by an Oklahoma City doctor and calling for one tablet a day on retiring. In a period of 92 days she purchased 180 tablets. A handwritten note found in her apartment shows she had made a $300 expenditure for some type of "dope" during one period. Her use of narcotic substances was further revealed by her attempt to commit suicide in September 1973. According to a statement by a friend, Connie Edwards, Karen called to say she had tried to kill herself by taking an overdose of drugs. Ms. Edwards states she went to Ms. Silkwood's apartment and found her "in a stoop on the sofa."

Her second problem was she had become contaminated by plutonium.

A lot has been made of this peculiar circumstance, with one

group maintaining she was deliberately exposed by a faction in the struggle between union and management, another espousing the theory that she deliberately contaminated herself, and a third viewing it as an attempt on her life by a group interested in preventing her from telling what she knew about the theft of the valuable but toxic plutonium from the Cimarron plant.

According to the evidence, Karen was involved in several separate incidents, starting on Tuesday, November 5, 1974. She had been absent from work the first four days of November. FBI documents contain a reference to a statement from Donald Gummow, a fellow employee and Karen's close personal friend, indicating she had spent those four nights with him at his residence. Gummow and Silkwood had received letters of reprimand from Kerr-McGee stemming from an incident on Thursday, October 31, when Karen had taken a prescription drug without informing her supervisor, as required by company regulations.

After reporting for work at about 1:20 P.M. on Tuesday, November 5, she dressed in a smock and performed routine paperwork in Room 135, a metallography laboratory. At about 2:45, before leaving the area for a break, she and her supervisor monitored themselves. A hot spot was found on her superior's plastic disposable shoe cover. A health physics technician was called. He monitored the floor but found no further sign of contamination. At about 3:15 P.M., Karen left the lab after checking herself thoroughly. She was apparently free of radioactivity at this time.

At 3:45, after returning, she dressed in a protective coverall and donned a pair of thin plastic gloves with tape around the top of the wrist, in preparation for working in a special closed container called a "glovebox." In this process, the technician is seated at a large tablelike device which has a glass top so the contents inside are in clear view. The workers insert their hands into special gloves fixed over holes in the side of the box allowing them to reach inside to manipulate samples without direct skin contact.

The work can be seen through the glass top but the radioactive material cannot get into the air. The thin plastic gloves, placed over the hands before reaching inside the box, provide a double layer of protection and serve as a backup if the outer, more rugged glove is damaged.

Which is exactly what seemed to have happened to Karen.

After each task is performed inside the glovebox, the workers, upon removal of their hands from the strong outer gloves and sleeves, check for contamination, using an instrument mounted on the front exterior of the box. Karen followed this procedure carefully and at about 6:30 P.M. found radioactivity on her fingers. She called another lab analyst and the health physics technician returned.

After following the prescribed cleanup process, Karen was again checked and the findings indicated she was free from the earlier reported problem.

According to W. J. Shelley, whose statement is included in the FBI review of the case, no leaks were found in the glovebox used by Silkwood, even though her fingers were contaminated.

As a precautionary measure, Karen was required to give urine and fecal samples for five days after her exposure. Close monitoring of her body samples, as reported in a paper by the union, shows that when she gave a urine or fecal sample at home, the level of contamination was high, and when a sample was taken under more controlled conditions in the plant, the levels were low. The AEC investigation concluded that plutonium had been added to at least two of her specimens after they had been voided.

Wayne Norwood, the Health Physics and Industrial Safety Manager of the Cimarron plant, also commented Karen Silkwood was the only individual in his twenty years of experience who, when found to be possibly contaminated, had been uncooperative in the submission of samples.

Another unexplained incident followed the one on Tuesday, November 5. During a routine test on Thursday, November 7, a nasal smear was taken from Karen and a high radiation count was present. She claimed this was from bringing up contaminated mucus from her lungs or stomach and that the exposure had occurred on Tuesday. But on November 5, Karen had been working on plutonium pellet lot number 35. The contamination in the smear was from another pellet lot, number 29.

Karen suffered from sinus trouble. According to a statement from Drew Stephens, she was a mouth breather. Ingestion, then, of a very small amount of plutonium is possible.

In any case, there is some confusion as to what really happened on November 5, 6, and 7 in regard to her contamination.

But on Thursday, November 7, a team of health physics technicians from Kerr-McGee accompanied Karen home and made a
survey of her apartment. They found significant levels of
plutonium in the bathroom around the toilet and in some lunch
meat and cheese in her refrigerator.

Her roommate, Sherri Ellis, was awakened and examined.
She was found to have slight contamination in two areas. Drew
Stephens, who had spent the night at the apartment but had left
early Thursday morning, was later checked. Reports show neither
he nor his own dwelling or clothes were contaminated.

In an effort to settle, once and for all, the level of Karen's
exposure, she was sent to the Los Alamos Scientific Laboratory for
examination. Included in these tests were Sherri Ellis and Drew
Stephens.

The results, given by Dr. George L. Voelz, indicate the
problem of contamination was so low as not to present "a significant health hazard . . . either now or in the future."

Even though the total amount of plutonium found contaminating the individuals and the apartment was very small, it
appears somehow, some way, someone removed a quantity of the
material from the Cimarron plant.

A review of the facts suggests these possibilities: Karen
Silkwood could have contaminated herself and added plutonium
to her own urine and fecal samples. Supportive of this theory are
statements made by her roommate, Sherri Ellis, to the effect that
Karen might have stolen a small amount of plutonium. She also
thought it most likely Karen had "spiked" her specimens to prove
her allegations about Kerr-McGee.

A motive is present. The intensity of her desire to embarrass
Kerr-McGee and gain publicity for the union cause is unquestion-
able. There is no doubt her contamination could have been used
to gain national awareness.

Against this theory, however, is Karen's well-known abhor-
rence of becoming contaminated. But she was aware of how to
handle plutonium and could have done it. Such an act is not
without similar self-sacrificial precedent in the long and tumultu-
ous history of disagreement between American labor and management.

Another possibility is that her friend Drew Stephens, with or
without her consent, managed to place plutonium in her apart-

100

ment and in several of her body samples. Unfortunately, Stephens did not testify before the House subcommittee, and we do not have his first-hand version of the story.

The facts involving Drew Stephens are taken from the records of the hearings before the Subcommittee on Energy and Environment of the Committee on Small Business in the House of Representatives.

First, he had access to Karen and her apartment. Second, at least one of the specimen bottles used by Karen at home, and from which a sample with a high radioactive count was taken, is said to have come from the trunk of Stephens's car. Third, Stephens had worked in the Cimarron plant and had been active in the union. He is the one who met the union representative, Wodka, at the airport on the evening Karen died. Fourth, Stephens, who was then working as an auto repairman, had the battered white Honda hauled away to Oklahoma City, where he held the vehicle while the union located an accident investigator to look into their suspicions of foul play.

According to a statement from an individual who saw and actually touched the left rear fender of the vehicle within hours after the accident, there was no damage to the auto body in that area. Two days later, after the car had been in Stephens's hands, the investigator hired by the union examined the same body panel and found damage.

It is possible this is a series of coincidences. Or, conversely, these facts may indicate Drew Stephens, working with the union, assisted them in a campaign aimed at Kerr-McGee, and used Karen as a means to get to them.

Members of the Kerr-McGee management team at the Cimarron plant must also fall under suspicion. The facility's safety record left something to be desired. And there is a documented incident of an employee tampering with a quality control photograph because he felt it was expected of him by his superiors, and he didn't want to have to make the photo over again. Added to this, there was a move to discredit and remove the union from the plant.

No part of a conjecture involving this group, however, stands up after reviewing the available evidence.

The AEC either knew of, or very soon discovered, the safety violations and the photo-negative doctoring. And the union won

101

its vote to remain at the plant. In short, there was no rational basis or motive for Kerr-McGee management to have become entangled in an attempt on Karen's life—unless they were involved in a plot to remove large quantities of plutonium from the Cimarron center for resale on the black or gray market.

Which brings us to the part of the story that has been given wide national publicity and has kept the Silkwood affair alive for more than four years.

This version holds that in her research to discover violations, Ms. Silkwood came across shocking news concerning the illicit removal of plutonium from the Kerr-McGee plant for resale on the international gray or black market. Such theft, depending on the version being told, was done either without Kerr-McGee's knowledge or with their full aid and consent. The missing manila envelope, in this theory, contained proof positive of theft and the identity of the individuals involved. If true, here indeed are grounds for killing. And drama: the alleged theft of a deadly, strategically important radioactive substance vital to the construction of atomic weapons, a lone girl struggling valiantly to get the information to the proper authorities, a vicious car chase, then finally murder. A great deal of work has been done on this thesis and some of it bears review.

A number of spectacular accusations, such as the report attributed to David Burnham, a *New York Times* reporter, about 60 pounds of plutonium missing from the site, have caused widespread comment.

Additional credence for this version of the story comes from the actions of an FBI agent, Larry Olson, who at the time was a forty-four-year-old, six-feet tall, slim, fair-haired man with fifteen years' experience on the force.

He came across the rumor there were large amounts of missing plutonium while investigating Ms. Silkwood's death and in an "official letterhead memo" seconded, according to published reports, by Ted Rosack, the acting agent in charge of the Oklahoma City FBI office, requested the opening of a new investigation into the matter.

The request, and a second official petition, were both refused by FBI headquarters because a study of the AEC records showed no such loss. A full inventory of the facility had been made, and although shortages were found, they were explainable as normal MUF—missing or unaccounted-for losses.

All this sounds mysterious, especially because of the controversy stemming from Karen Silkwood's death. But an objective view of the facts as presented to the House investigating subcommittee tends, regrettably for those who love mystery, collusion, and meetings in dark places, to show the incident for about what it was.

Ms. Silkwood displayed signs of emotional disturbance and had espoused a cause with more than average vehemence.

During the thirty-one month period, March 1972 to September 1974, nine inspections and one investigation were conducted of the Kerr-McGee Cimarron facility. As a result of these inspections and investigations, nineteen items of noncompliance with AEC regulatory requirements were found. And two letters from NRC to Kerr-McGee expressed concern regarding excessive MUF rates.

In the atomic industry, loss is an expected event. In any chemical process where many thousands of pounds of a material are run through, certain amounts are missing at the far end. The material does not vanish but, rather, is accumulated in pipe bends, tiny surface irregularities on the inside of the processing tanks, and in other areas where the chemicals are suspended in some form of solution.

Kerr-McGee's plutonium MUF was higher than expected, and at least some of this loss might be accounted for by the removal of tiny amounts of plutonium from the plant site by employees.

But the total loss was sufficiently small to be explained away to the AEC. In other words, the amounts were not too far above the loss levels predicted for the operation of the processing system.

The real refutation of this version of the story comes from another source. As noted earlier in September 1976, the General Accounting Office (GAO) submitted a report indicating that as much as 11,000 pounds, mostly plutonium and enriched uranium, was unaccounted for at government-owned facilities around the country. Another document, not released, showed an additional 15,000 to 20,000 pounds missing from privately owned facilities. These figures are interesting because an

August 1977 update indicated the loss from nuclear facilities in the United States to be about 8,000 pounds.

But neither figure, according to the reports, also mentioned previously, is correct, as both sources deliberately left out the MUF from the two major U.S. weapons facilities in Colorado and Tennessee for fear of providing classified information to unfriendly intelligence services.

The reported figures did not include the additional amounts of low-grade uranium missing from the 65 atomic reactors operating at that time to produce electricity for various parts of the U.S.

The total MUF from the Kerr-McGee facility was included, however, and no special mention was made of any astounding shortage even though one report did cite other locations where large amounts of material were unaccounted for.

The most famous case of MUFfed enriched uranium, and the one which has given rise to the idea of an international black or gray market, concerns the Apollo, Pennsylvania, facility. The AEC, FBI, and CIA were called in to investigate national and international aspects of a twenty-year loss of 381.6 pounds of the valuable element. One of the facility's top executives, according to The New York Times, was suspected of being an agent of a foreign government, and there was widespread speculations the lost material had been sold to or stolen by Israeli agents.

The AEC, responsible for the facility, finally concluded after much study that there was no evidence of any theft and the missing materials were lost due to crude statistical and measuring systems working in combination with processing loss. What no one seems to have focused on in their alarm at these figures is the acceptance of loss due to processing. The Kerr-McGee Cimarron operation had, for example, an estimated allowable MUF of around 1.8 kilograms (about 4 pounds) in a given time period. This was increased when several months of operation indicated their loss would be higher. The Kerr-McGee plant, like virtually every other nuclear installation in America handling enriched uranium or plutonium, had a minor problem of accountability with the AEC and received numerous memos on the subject. But no action was taken and no censure was imposed because everyone involved knew and understood processing-loss accountability was not a completely refined science as applied to atomic materials. No matter how hard any group of engineers might try, the amount of

104

material remaining in the system, trapped in the piping or thrown out with other low-grade waste, could not be pinpointed by preproduction-run estimates.

Sixty pounds of plutonium was not found to be missing from Cimarron, even though unaccounted-for amounts were a few pounds greater than the levels set by a preoperational estimate. The implication of large losses and Karen Silkwood's involvement in the matter, either as detector or thief, is difficult to sustain in the face of the plant's not having had documented major losses beyond the limits of measurement error.

Another problem with the theory that she found evidence of large-scale smuggling is the fact she did not mention her findings to any of her friends. Indeed, on the night she died, just minutes before leaving the Hub Café she made a telephone call to be certain her meeting was set, but made no reported mention of such a blockbuster fact. And earlier that same day, in a meeting with an AEC representative to discuss her contamination, she did not bring it up.

From all accounts Karen Silkwood was not one to quietly keep a secret and she seems to have shared her innermost feelings with her friends.

Conversely, if she had been engaged as one of the thieves, she would have had to been a part of a very long chain of people. Plutonium is not the easiest material in the world to sell. A letter to a foreign nation or a clandestine contact arranged by an unknown person, offering a quantity of the element at some price, would be treated with great suspicion and probably be reported to our own FBI.

The implications of a huge loss from a nuclear facility go even further than this.

Such a quantity would be very valuable. If the loss occurred without management's involvement at the very highest levels, the resulting reduction of corporate profits would have certainly brought strong investigative measures from the company itself. From this, we may deduce no large-scale theft would be possible without the collusion of top management. No evidence has ever been offered to indicate this as even a remote possibility.

Actually, much of the theft theory stems from a combination of incidents. Sherri Ellis, Ms. Silkwood's roommate, is said to have been the originator of the idea Karen might have smuggled small quantities of plutonium from the plant; the apartment occupied by the two girls was found to be contaminated; analysis indicated the radioactive material had come from the Kerr-McGee facility.

These facts, supported by rumors and the odd behavior of a number of people, one of whom insisted on leaving clues for the FBI in public telephone booths and other spylike places, became distorted. Fertilized by the desire of a few individuals to see Kerr-McGee placed in an embarrassing public light, fiction was embraced as fact.

The concept of dramatic losses and a theft-smuggling ring is interesting and exciting as speculation, but seems to be on imagination and wishful thinking as opposed to reality. While it is possible to build up the theory of an elaborate conspiracy, it has to encompass a vast number of responsible people, including several Congressmen.

All this leaves one unanswered problem. What about the mysterious manila or brown envelope, sworn by a single witness to have been in Ms. Silkwood's possession shortly before the accident (or, if you will, murder)? It supposedly contained information from her hours of interviewing Kerr-McGee employees and searching through company files.

The envelope, according to most versions of the story, was missing when the wreck was found. Karen Silkwood's possession of the envelope is given by some as the motive for her death. Their contention is a party or parties unknown ran her off the road, she died in the crash, and her assailants made off with the documentation of Kerr-McGee's wrongdoing.

But in the light of the subsequent AEC investigation, in which the same files Ms. Silkwood used were reviewed and many of the people who had talked with her offered testimony, the evidence that she was killed to prevent information from reaching the press and the AEC seems a little thin.

Only one witness has sworn, in an affidavit, that Karen had such an envelope in her possession at the Hub Café. Numbers of other union co-workers, however, who met with her the same night and spent time in her company, are not sure, and none have stated they recall seeing such an envelope.

Papers were found in the car and around the scene of the wreck. They were boxed up and turned over to Drew Stephens, who had Karen's parents' permission to claim the effects of their deceased daughter. The envelope, according to Stephens, was not included. An FBI investigation to determine if the envelope ever really existed proved indecisive. But aside from the sole witness, neither Karen's roommate nor her friend Stephens testified under oath as to its existence. And the one person who stated the envelope did exist did not appear at any point in the hearings.

In addition, the single testimony concerning the envelope was taken by the union, not the FBI. And the OCAW can hardly be described as being impartial in the matter.

All this leads inevitably to a single conclusion. The only mystery in the unfortunate girl's death is why some individuals have chosen to ignore both fact and logic to develop a series of theories based on speculations of sinister conspiracy.

There is a part in all of us which loves the mysterious and longs for tales of dark-of-the-moon drama. And sometimes we let this portion of our hearts rule our heads. A lot of the people responsible for the Karen Silkwood mystery have done exactly this. There is no reasoning with them, especially when their assertions of intrigue are closely joined to claims against the nuclear industry. The anti-atomic forces have taken this incident and expanded it all out of proportion to serve their needs. In short, enough people want the shadow of a conspiracy for there to be one.

Regrettable in this instance is the fact that the darkness cast by the suppositions about Ms. Silkwood serves to obliterate the deeper insight we could obtain into the inner workings of a modern nuclear manufacturing and processing site. And that insight is disturbing.

To a large extent, we see a somewhat primitive operation, in part due to the cost difference between a chemical processing plant and an installation for the handling of atomic materials. In today's marketplace a first-class chemical plant, with all necessary pollution and worker-safety equipment, can cost several hundred million dollars. A plutonium-processing plant can be built for a far lesser amount.

With great sums of money at stake, the design factors of the chemical plant will be centered on smooth, uninterrupted operation. Automation is at a maximum and manpower is used to service the equipment, not forward the processing itself.

In the plutonium plant, the opposite is true. Manpower is the basic force in the system. A multitude of jobs are performed by humans. The difference is more than philosophical.

People are allowed to come into close contact with radioactive substances instead of remaining protected from the flow. And since the dollar investment is far less, the continued throughput of the facility, which assures its maximum usage and therefore the highest possible dollar return, is not as vitally important.

The Cimarron plant was, from all reports, an installation as advanced as the state of the art allowed. But the art needs to advance a great deal more if we are to have safe handling of nuclear substances.

By the most recent testimony, there have been MUF losses of thousands of pounds of toxic substances with long half-lives, such as high-grade uranium and plutonium. It is not enough to demonstrate some kind of international smuggling or black market, but it is a sufficient amount to cause real concern when the damage potential of radioactive materials is considered.

In short, less care appears to be exercised in the processing of highly enriched uranium or plutonium than in the smelting of gold, where loss factors in the hundreds of pounds would be considered totally inexcusable. In fact, the gold industry has become so refined as to be able to measure its MUF quantities in troy ounces.

The upshot of the matter is clear. We have embarked on a manufacturing system too directly related to the laboratory technology from which it sprang. No one has given enough engineering thought to the design of a method of processing radioactive materials. Techniques in other industries are far ahead of those apparently being used in working with plutonium.

The gap can be closed, but only by large investments by responsible firms with proven backgrounds in this vital area.

If nothing else is derived from the Karen Silkwood matter, an understanding of the risky production techniques in use at the time should stand out as a warning. With 73 people exposed to internal dosages of plutonium during the operating history of the Cimarron plant, the need for better, more efficient processing is made alarmingly clear.

The most important question brought out by the death of Karen Silkwood is not whether she was murdered, but how much will have to be done to make the refining of atomic materials as safe as any other major manufacturing operation. Procedures acceptable in the time of the Industrial Revolution are no longer tolerated by reasonable people. The message to the atomic processing industry is clear: they must advance, and quickly.

To update the Silkwood situation: On November 5, 1976, Karen Silkwood's family filed a $160,000 law suit in Oklahoma, charging officials of the Kerr-McGee corporation with conspiring to prevent Ms. Silkwood from organizing a union and reporting nuclear safety hazards to the U.S. Government. The civil action further alleges corporation officials, a former newspaper reporter who cooperated with the FBI, and three agents conspired to suppress information about the intimidation and harassment of Ms. Silkwood. Finally, the claim charges Kerr-McGee officials with failure to exercise sufficient security controls over plutonium produced in their Cimarron plant in the incident in 1974 when Ms. Silkwood and her apartment were contaminated.

This suit, should it survive anticipated legal challenges, might end any remaining mystery. All sides will be able to come forth in a court of law, where there are strict penalties for perjury, to state their cases. It will probably be a good thing, but there is no indication the antinuclear forces would be any more inclined to accept information presented in this trial than they were to accept the facts presented in the House Subcommittee's report. To them, Karen Silkwood is a nuclear martyr. Her death has enveloped her work and actions in an emotional aura. Those who believe she was killed by persons unknown to prevent her delivering incriminating documents to the government, or because she knew of a massive plutonium smuggling operation, will never accept any other explanation.

■ Plutonium and enriched uranium have been stolen from nuclear processing plants in the United States. There is no real question of this fact. In the Karen Silkwood matter, for example, someone illicitly removed small amounts of plutonium from the Kerr-McGee plant, or the Silkwood apartment would never have been contaminated.

The question is not whether there have been thefts but how much has been taken. And what materials were involved.

During the Dingell hearings, a joint report from the Energy Research and Development Agency (ERDA) and the Nuclear Regulatory Commission (NRC), stated there had been no large-scale thefts of plutonium or enriched uranium. This was severely questioned. But Robert W. Fri, acting administrator of ERDA, who testified that full investigation of every suspicious incident had led to the conclusion: "There is no evidence that a significant amount of special nuclear material has been stolen or diverted."

Immediately after the meeting, however, Michael Ward, the staff director of a House investigation into the effectiveness of federal measures to control special nuclear materials, made a statement of his own. In part, he said United States intelligence officials had "strong suspicions" highly enriched uranium, a material which could be used in nuclear weapons, was stolen from a U.S. atomic facility in 1965.

Rumors of this kind have been circulating around Washington for a number of years but no one has come up with proof. Another prevalent theory in some areas is the CIA has been involved in passing critical atomic materials to selected governments on a *sub rosa* basis.

There is no question about the purchase of certain nuclear

elements being a matter of great political concern. The recent disclosures of Israeli clandestine operations to secretly land two hundred tons of uranium concentrate into their country shows how far a government will go to keep knowledge of its nuclear business from other nations it considers hostile.

The Israeli story is a good one, filled with ships sailing at night to deep ocean rendezvous, machinations as governments cover their tracks, secret agents, and the trading of nuclear secrets.

In brief, in 1967 and 1968 the Israelis had a need for a large quantity of enriched uranium and plutonium to manufacture nuclear explosive devices at their French-built reactor station south of Haifa at Dimona in the Negev Desert. Fearing an open market purchase might push the Soviets into furnishing nuclear armaments to the Arab states, an elaborate scheme was developed. The uranium concentrate, commercially called yellowcake, was purchased through a defunct West German chemicals firm after assurances were given by the coalition government of Christian Democratic Chancellor Kurt-Georg Kiesinger that the transaction would be allowed to appear as a matter between two private companies.

Informed sources indicate in exchange for their aid, the West Germans received an advanced uranium separation process developed by the Israelis and vital to the production of atomic weapons.

Once the uranium concentrate was purchased, it was loaded onto a ship, the Scheersberg A, headed for Italy. But the vessel never reached its declared destination of Genoa, arriving instead in Iskenderun, Turkey, minus its cargo. A certain amount of attention in the European Community was directed to this mystery, but it took months for active investigation to begin. By then, the cover-up was tightly in place. What happened is that the ship sailed from Antwerp with the cargo of uranium worth about $3.7 million, bypassed its stated port of Genoa, went on radio silence, and in the dark of night a hundred or so miles from Haifa in the Mediterranean Sea, met with an Israeli vessel. There, working in pitch blackness and protected by the presence of two Israeli gunboats, the crews of the two vessels transferred the cargo. Once it was safely aboard the new ship, Scheersberg A sailed quietly away, and days later arrived in Turkey.

The secrecy of the operation was sufficient to prevent its

becoming known to the Soviets, as well as to most Western nations until several years afterward. Even today, there is some uncertainty as to the exact details.

Many nations find themselves in situations similar to the Israelis in 1968. A major movement of certain atomic materials required in the development or production of nuclear weapons is bound to cause response from other nations interested in maintaining a status quo. The efforts of the white South Africans in 1977 to convert a nuclear power station into a weapons arsenal and manufacture a bomb is another excellent example. Both the Soviet and U.S. governments, joined by the other nations with the capability of making nuclear explosive devices, moved to halt this attempt with a swiftness uncharacteristic of government dealings. The notable exception to the list of participating nations was China, which stayed out partially in respect to the wishes of the other countries and partially in response to the counter action having been instigated by the Soviet Union.

With nations going so far to control access to information about their purchases, and conducting nuclear experimentation in secret, it is no wonder there is a problem in determining whether or not large amounts of enriched uranium or plutonium have been surreptitiously removed from U.S. facilities, where controls are tighter than in other countries. But what about the nuclear reactor installations of other nations? In comparison with what is known about U.S. shortages, our knowledge of overseas MUFs is nil.

It has long been known that shortages have occurred in India, and it is suspected the government or at least highly placed members of the government have been involved in the sale of radioactive raw material on the world open market. Records show this activity has been conducted without official Indian sanction and the profits from the sale have gone into the pockets of a few individuals.

Aside from this incident, and one or two other minor happenings, nothing concrete is available on the state of nuclear materials theft in the world. But because of technical problems it cannot be too widespread, and while potentially profitable, the number of transactions would make this a poorly paying activity for any small organization.

As differentiated from other kinds of theft, the real work of

113

the nuclear thief starts after he actually has the radioactive substance in his hands. If the element or elements are not properly stored and great care used in their transportation, they will injure or maim all connected with their movement. Unless caution is used in storage they will burst into flame and grow hot enough to vaporize, sending out plumes of radioactivity.

It would take a knowledgeable, well-planned effort not only to steal, but to handle the contraband, and a team of experts would have to be along in order to look after the shipment. Some of the most desirable elements are the most volatile. An entire branch of science has even developed to deal with the delicate problems associated with plutonium and enriched uranium. Specially designed canisters, built to a critical geometric formula, must be kept separated by precise amounts of space and never stacked over certain heights.

The antinuclear faction has called attention to the dangers inherent in the possibility of theft of atomic materials, especially by a terrorist organization, but to date, no evidence exists of this type of incident. Just because it hasn't happened yet is no indication it will not happen in the future. In fact, as more and more atomic facilities are built, the chance becomes greater. And as installations crop up in smaller countries, notably some of the African nations, the probability of a government's making a clandestine deal for a quantity of plutonium is very high—especially if we accept today's standards of materials accountability. After a period of a few years, a sufficient amount of allowable MUF would build up, and could be used to cover a large transaction.

The reason a quasi-military organization would want to acquire the elements would be to develop a weapon. There has been a considerable amount of Sunday-supplement publicity about how easy it would be to construct an atomic bomb, and from a casual reading it would appear any individual could make one in a home workshop. But although the principles are simple, the construction, aside from theoretical considerations, is complex and time-consuming.

A news story in 1977 about a Princeton University student, John Phillips, describing a paper he had written on how to construct a nuclear explosive device by acquiring $150,000 worth of plutonium and a few thousand dollars' worth of other equipment

114

is a case in point. Phillips received national publicity and attention for his efforts, which were ingenious. But his work did not deal with the full array of practical difficulties which come into play after all the materials are in hand. The South African attempt to secretly construct such a device shows how well things are being monitored and how very difficult it is to go from paper theory to real explosive hardware. The actual stages of construction and the handling of radioactive materials pose problems which do not lend themselves to easy or inexpensive solutions.

This is not to imply it would be impossible for a group of well-financed, strongly dedicated individuals to make a crude bomb. Even the staunchest, most positive supporters of nuclear power do not assert this. But again, it's a very unlikely occurrence on today's international scene, due to the climate of anxiety and suspicion. Furthermore, the group would have a weapon of unknown performance and reliability. It might not go off at all, or it might explode with more force than anyone expected.

It is a little hard to imagine a terrorist group getting together the expertise and financing for such a project, when the same amount of money might be better expended in obtaining a ready-made weapon by a sudden shock attack on an ill-defended arsenal. There are several of those in the world, including some controlled by the U.S. Armed Forces.

The construction of an illicit explosive device, then, will probably come from a government, or a government-backed group, capable of buying the necessary materials secretly to avoid intervention by the present nuclear powers.

But is it necessary to have an atomic bomb in order to influence a particular situation? Is there a nuclear weapon easier to manufacture yet still impressive enough to be intimidating? A weapon which might strike even more fear into a society, achieving the terrorist group's aim of publicity and public awareness?

The answer, sadly, is yes. There is, in fact, an entire class of such weapons. And while they do not possess the destructive force of a nuclear bomb, they have an inherent fear quotient of a far higher level.

While admittedly difficult, it is within the realm of possibility for an organization to acquire a limited amount of plutonium, radioactive cobalt, or other "hot" material. This could be achieved by theft, secret international purchase, or through a direct gift from a supportive government which possesses a reactor.

In the last example, the reactor itself might be used to irradiate, or make radioactive, some available substance. The half-life of the material, as long as it was a week or so, would be unimportant, as plans could be made to use it before it became too cold to injure or kill.

Another source of noncontrolled material might be industries which use radioactive substances for quality-control measurements and other commercial activities.

Once obtained, such material could quickly and simply be made into a weapon of respectable proportions. A limited amount of chemical explosive—even one as common as dynamite, which would work but would be less suited to the task than, say, one of the "plastic" concoctions—connected to an electric or clock-work timer and primer, also easily available or quickly fashioned, is all that would be needed to complete the device.

With the radioactive material encased in a fractionable container along with the chemical explosive, and the timing device set to a specific ignition point, the weapon—which would be about as small as a light suitcase—could be left in a variety of places. It might be slipped into an air-conditioning duct, for instance, or placed in a public locker of an air terminal, or simply stood innocently against a wall. Great size is not a requirement for an effective bomb. A small one would, in many situations, be as efficient as a larger device.

When the timer sets off the chemical explosive, phenomenal heat is generated. The protective lead container is vaporized, as is much of the radioactive material, and the gaseous mass is forcibly dispersed over a large area. If an air-conditioning system were chosen as the site for the device, it would be possible to design a container shaped to direct the force of the blast down the duct into the heart of the unit, where fans would force the radioactive vapor and particles throughout a building in a matter of minutes.

Anyone who breathed the hot material might be in trouble. The panic potential would be tremendous—as would, of course, be the publicity value.

Radioactive materials have one interesting thing in common. They can be detected in a number of electrical or mechanical ways, but without special apparatus, there is no way to know of their presence. If, after the blast, an announcement of the nuclear nature of the explosion was made public and trained contamination crews were sent to analyze the radioactive levels, the anguish of the people present would be enormous. The shock effect of the weapon, coupled with the threat of its future use, would cause long-term panic and disruption regardless of what authorities might say about its relative lack of effectiveness or danger potential.

In short, those individuals who even imagined themselves contaminated by the blast would react fearfully. But there would be no way to tell. A few, who might have received high-dosage exposure, could show symptoms in a day or so. But just how serious the level of contamination might be in any given person could only be determined by an extensive medical examination. And even then the gnawing fear of a developing cancer would produce anxiety.

Unconfirmed reports indicate a terrorist group in Europe has actually constructed and planted a device of this type. It was found before the built-in timer detonated the chemical explosive, so no one was injured.

The idea for such a weapon is not new. It was discussed by our own military and has been the subject of speculation in a number of technical articles. A little work by a small number of informed people would produce an effective prototype. It could be done even more quickly if a few fanatics could be found to handle the hot materials with only minimum protection. Such a step would probably result in the deaths of the people so engaged, but organizations devoted to terrorist actions have come up with suicide squads on a disturbingly frequent basis.

To date, no weapon of this type has gone off in an inhabited place. It is hoped none ever will. But if it should we will quickly find how unprepared we are to withstand nuclear emergencies and how limited our resources are on the local level to deal with an atomic incident.

A review of the capabilities of police and fire departments in a number of major U.S. cities, including New York, Chicago, Los Angeles, Dallas, Houston, and others, indicates low levels of preparedness for dealing with large-area contamination or sizable numbers of people exposed to radioactivity.

The states are better off in this regard because they have access to the National Guard and other military units which have received special training, but the level of health physics' capabilities is not generally very high.

Doctors and emergency medical care facilities are also not sufficiently prepared to cope with this kind of a problem. To date, there is no reason they should be. No widespread exposure has occurred, and there are numbers of experts on terrorist actions who maintain a nuclear device of any kind is beyond the scope and desire of the individuals who compose the action forces of insurgent groups.

Fortunately, the right elements have not come together to cause such an incident. They may never do so. But the possibility does exist. And it grows as radioactive material becomes more widely distributed and available throughout the world.

■ There is little doubt the worst nuclear accident to date in a Western nation happened in 1957 in England. The Windscale incident, as it has come to be known, is similar in many respects to the Soviet disaster of the same decade. The results in terms of injury, loss of life, and property damage were, however, significantly less in the English event, as no individual not directly connected with the operation of the facility was physically injured.

The Windscale No. 1 pile is a graphite-based plutonium producer. Along with its nearby neighbor, the Calder Hall nuclear facility, it composed a major portion of the United Kingdom's atomic research and development capability. The area has been likened to our own Oak Ridge, Tennessee.

The antinuclear energy movement has publicized the Windscale matter, pointing to it as a prime example of how close humanity has come to a real disaster. The pronuclear forces have tried to ignore it. But if anything was proved by the event it was that we do tend to function at the extreme limits of our knowledge of nuclear science. And if facility operators are presented with the choice between a full meltdown or the release of radioactivity into the environment, the release will occur. It also demonstrates how even unanticipated difficulties can be overcome by the courage and coolness of those who run the installations.

The Windscale incident is an interesting story. The press at the time vacillated in its reportage, some papers hewing to the line of the official plant release and maintaining everything was well; others using boldface black type to silently scream doom, and for weeks afterwards playing up the event as its full international implications unfolded.

In the end, after the committees had met and filed their

voluminous reports, things went back to normal. But the No. 1 pile at Windscale was out of service. Ten years would pass before the radioactive remains could be inspected closely by man.

Forewarning of the accident occurred in September of 1952, when the unit spontaneously released a small blast of heat and radiation during a phase in which it was shut down. Temperatures rose dramatically in the graphite material which encased the uranium in the core, but the radiation was mostly contained and none of the uranium cartridges caught fire. A study of the surprising incident revealed that graphite, used to slow the neutrons to give them a better chance of hitting other uranium nuclei, built up and held energy after being bombarded by neutrons. First explained by the Nobel Prize-winning physicist Eugene Paul Wigner, this form of buildup bears his name and is known worldwide as "Wigner energy."

The results of the study called for a planned, controlled reduction of Wigner energy in the pile on a regular basis to prevent an unexpected and unplanned-for spontaneous release. By the end of 1956, this had been done on eight occasions.

The procedure was simple. The pile was shut down, a series of instruments to measure heat and radioactivity were arranged in appropriate places, and the rapid airflow, which took the place of a water cooling system in this particular reactor, was shut off. The result was an almost instantaneous increase in the temperature of the uranium and graphite which made up the pile. This added temperature started a slow release of the stored up Wigner energy in the graphite. Once initiated, the process became self-sustaining. At a given point, when it was determined enough energy had been released, the fans were restarted and the pile was brought back into operation.

The trick, it appears in retrospect, was to obtain an even and equal release of all Wigner energy. On three occasions prior to the one in which disaster struck, in order to free all the pent up energy and get proper instrument readings it had been necessary to heat the pile twice. This appeared to solve the problem, so the technique became a normal part of the system operation. On the day of the accident, the procedure was instigated in the usual manner. The pile was shut down, the main fans stopped, and after checking all instrumentation and verifying the pile was indeed off, the technicians waited. At the prescribed time, readings were taken and the decision was made to repeat the process to complete the release.

A later review of the data indicates this is where the mistake occurred. The physicist in charge decided to reheat the pile to boost the release rate. But he made his decision without sufficiently detailed instructions. The Pile Operating Manual had no special section on how to achieve the Wigner release, and it was left to the judgment of the individual operator as to how long and how hot the pile should be allowed to become. The outcome was disastrous—and it spread the name Windscale around the world.

■ John Bateman was an early riser. As a cattle breeder and farmer, he was accustomed to the cold, blustery dawn hours when darkness still covered the rolling, green countryside. He stood outside his main house looking toward the horizon where the cooling towers and spires of the Calder Hall nuclear power station and the Windscale plutonium refining installation made blacker shapes against the brightening sky. The man-made structures were the first changes in the landscape to have occurred in years. Because it was too far north to attract the industrial developers who bought and built in the more mechanized south, Bateman liked his land and his little town of Yottenfews. "Town," really, was too grand a word for the small collection of buildings and homes. Three families lived there, more than a mile off the main highway. It was quiet on the coast of the Irish Sea and life had gone on in an unchanging fashion for generations.

But somehow the atomic plants had altered something. His neighbor, Hewitson, the dairy farmer, had filed a formal complaint when the Government put a high wire fence across one of his meadows. But it availed him of naught. Even the Ministry of Agriculture remained deaf to his charges that the new facility had affected the fertility of his cattle.

The matter was hard to judge. Officials had talked to him on several occasions but nothing was changed. Bateman shrugged and turned away from the vision on the horizon. He had work to do and standing around mooning over the way things were wasn't going to get it done.

Three miles away, a handful of technicians also had work to

do. Inside the Windscale enclosure, the No. 1 pile had been shut down at 1:13 A.M. in preparation for a routine maintenance release of the built-up Wigner energy. The men, acting as a part of a larger group, were verifying the closing of the system and replacing all defective thermocouples, the necessary devices for measuring differences in temperature.

They labored with almost exaggerated care, checking each output signal against the main control panel in the command room. Their team, a part of the 11:00 to 7:00 shift, had been through the Wigner release operation before and the work had become routine.

Finally, by 7:00 that evening, Monday, October 7, preparations were complete for the release maneuver. The graphite control rods were moved and the pile was actually opened twenty-five minutes later. Everything seemed in order. The physicist in charge gave his okay to the procedure, and the nuclear heating of the graphite continued throughout the night. Early on the morning of Tuesday, October 8, the heat flow was stilled and readings, taken during the process, were evaluated.

Several hours passed to allow the thermocouples to trace the temperature trends in the material they were sensing. The data produced considerable disagreement. Instrument readings were conflicting. Some of the controllers on duty believed the overall pile was cooling off. Others argued only certain portions of the graphite were cooling, thus showing no sign of energy release. A lengthy debate ensued and the final decision was to apply further nuclear heating to enhance the Wigner release process. The physicist in charge of the operation had no previous experience with the maneuver and none of his manuals outlined the technique. But based on his judgment, the overall indication was that insufficient heat had been applied to trigger the desired result. So, at 11:05 Tuesday morning, October 8, the pile was once again heated.

Thermocouples attached to the uranium cartridges immediately showed a dramatic temperature increase. They remained borderline safe for the next fifteen minutes.

The physicist monitoring the records was careful in taking readings but not astoundingly swift in his calculations. While the recordings of the temperature show not a single reading exceeded

122

the specified maximum, the actual heat must have been many times higher or else the buildup would not have been registered so quickly. Also, unknown to the physicist in charge, the thermocouples monitoring the uranium cartridge temperatures were placed to show the hottest or worst points during normal operations. During the Wigner release maneuver, however, the areas of maximum heat and stress were elsewhere in the pile. The result was a drastic overheating of some of the uranium.

Disturbed by the rapid temperature rise, the physicist ordered the control rods pushed in to allow the fuel cartridges to cool. But the damage was already done. During the minutes of highest stress, several of them had burst from the strain.

The result took time to reach its maximum potential. Behind concrete and lead shields in the heart of the radioactive pile, the Wigner energy release was still taking place. The graphite control sections grew increasingly hot. The uranium, in the split containers, was exposed to this heat and began to burn and oxidize. As uranium combines with oxygen it rusts in a fashion similar to iron. The major difference is uranium continues to heat until it is too hot to exist in its present state and turns to a yellowish-brown powder. All day on Wednesday, October 9, the uranium smoldered, giving out great amounts of heat not registered on the instruments in the control room because of the placement of the temperature measuring devices. This constant energy release caused the failure of other cartridges, adding more uranium fuel to the fiery mass.

During the next twenty-four hours, the fire slowly spread until it was affecting a major segment of the reactor pile.

Readings taken on Wednesday, October 9, indicated a continuing, expected, slow temperature rise in the graphite, which showed the desired release was taking place on schedule. One reading, however, thought by some to be due to a faulty instrument, showed a marked increase. By 10:00 P.M. it had reached a point where the physicist in charge was required, according to a specific written instruction, to take certain operational measures to cool the mass down.

Windscale No. 1 was of a unique design, no longer in modern use. Instead of water as the cooling medium, a series of chimneys was utilized to conduct airflow through and around crucial components. As long as the uranium remained sealed in its protective cartridges, the circulating air could be filtered to removal residual radioactive materials before being sent out from the towers in a reasonably safe, noncontaminated state.

The physicist in charge, following standing orders, called for the fans to be turned on. This allowed airflow through the pile to lower its temperature. At 10:15 Wednesday night his order was carried out and air streamed through the central chimney for fifteen minutes, drawing uranium oxide and other radioactive particles from the pile's core up the chimney and out into the open air, where they spewed in a long, hot, invisible plume in the direction of the prevailing winds. Readings showed the pile to be cooling, but the decision was made to give the pile another shot of air for safety. So again, at 12:01 A.M., Thursday, October 10, it received an additional ten minutes of cooling. Still not enough. Two more doses were called for, and the fans pushed hot material into the air for forty-three minutes during the damper openings which occurred at 2:15 and 5:10 A.M.

Later testimony indicates no one in charge of the operation had any idea there was a major emission of radioactivity during the first three air flushes. But alarm bells began to ring about thirty minutes into the fourth and final cooling effort.

The single graphite temperature reading which had steadily risen and set off the secondary cooling operation remained stabilized even though the pile itself began to lose heat early in the process. Then the pile radiation meter, located near the top of the tall air chimney or stack, showed a sharp increase in radioactivity. The physicist in charge noted the high levels but thought they were only a result of air movement through the huge vent.

At 5:40 Thursday morning the readings were still high. Then they seemed to fall for a few minutes. But by 8:10 A.M. they again showed steady increases. The situation was getting seriously out of hand.

The pile was shut down but was still releasing radioactivity—enough to cause the instruments on the roof of the meteorological building to show very high counts.

An immediate council was called. Tom Hughes, deputy works manager; Ron Gausden, reactor manager; and Tom Tuohy, the general works manager, agreed to ask for air samples from the health and physics staff. A quick study was undertaken and the

124

worst confirmed. Radioactivity in the air was up many times greater than normal.

The three managers called for half-hourly air testing at between ten and fifteen sites within the compound, then sat down to decide what course of action to pursue. Their meeting was of short duration interrupted by the news of another sharp increase in the graphite temperatures.

They were now faced with a difficult problem. But it had only one answer. The pile, composed of graphite shielding, could not be allowed to reach too high a temperature or a meltdown would occur. The release of nuclear radiation would multiply a thousandfold. So, after notifying their superiors, they took the only course available. They ordered the vents opened again at Thursday noon for fifteen minutes and used the fans at full blast to cool down the smoldering monster.

The radioactivity-measuring meter at the top of the stack went wild as soon as air started passing through, and everyone knew hot materials were being dumped into the atmosphere. The small group of technicians waited tensely in the main command room, scanning the temperature gauges. At the first indication of a remission of the heating they would order the flow stopped. Minutes ticked by but there was still no reliable sign of heat abatement. Then, slowly, one by one, the upward progress of the indicator needles on the instruments halted. The technicians immediately gave the order to stop the air flow through the stack.

But it was too soon. By 1:40 P.M., heat buildup was almost as bad as it had been earlier, and additional amounts of heavy radiation had streamed from the top of the tall chimney into the atmosphere.

Against their best moral judgment, but in an effort to keep the pile from melting, a final cooling flush was ordered and another five-minute flow called for. The gauges indicating radioactivity were well into their red zone-critical areas as they measured the outflowing air. Through it all, and working under conditions of almost unbelievable pressure, the management group battled to determine the cause of the malfunction. The health physics manager was given the job of finding out where and how far afield the radioactivity was going, and the remainder of the men wrestled with the technical problem.

They finally decided they were faced with one or more burst cartridges. Their analysis was correct.

They needed to examine the insides of the core, locate the ruptured units, determine how many there were, then in one maneuver quench the fire and dump the broken canisters into a protective container.

It was a difficult undertaking and had never before been carried out. The operation started going wrong from the first. Thursday at 2:30 P.M., the turbo-exhaust system was switched on to clear the air in the pile so an inspection by means of a built-in radiation scanner could be made. The pile was flushed clean and then, to everyone's horror, it was discovered the intense heat had jammed the gear drive of the mobile scanning apparatus so it could not be moved. Tom Hughes and Kenneth Ross, the national operations director, worked on the large machine, but it was no use. In a later testimony the pile manager indicated loss of the instrument was normal during a Wigner release; in fact, he said, even though the scanning device was in operating condition, just having been worked on the day before and left operational, it had become immobile at the end of the previous energy release maneuver some several weeks earlier.

It was at this time the health physics officers ordered air sampling trucks sent out into the countryside to map and quantify the amount of radiation which had spewed from the chimneys. The vehicles, moving slowly through the misty, bright green, rolling land stopped again and again as samples were taken and readings made. An elaborate map was started, but hours would pass before there would be a full understanding of how massive an amount of radioactivity was really involved.

Back at the plant, a special team was gathered. Tom Hughes and Ron Gausden, the reactor manager, dressed in their protective clothing in silence. Every man present knew what they had to do next was extremely dangerous.

Completely covered, wearing gloves, overshoes, helmets and respirators, the pair was going to visually examine the inside of the pile where the temperature readings were highest. In so doing, they would expose themselves to very large amounts of radiation.

The decision to risk station personnel had not been arrived at lightly. Tom Tuohy, the works general manager, after being informed of the reasonable expectation of having a burst cartridge, suggested an alternative scanning method. It worked, but its reading indicated almost impossibly high levels of radiation and provided the staff with little new information.

Ready for their task, the two men moved swiftly out of the main building to the concrete structure which contained the runaway nuclear pile. No one had ever attempted the maneuver they were about to perform, but according to the best available estimates, it was possible. However the potential for death or injury was very high.

The procedure had directness in its favor.

First, an air count for radioactivity would be taken to assure there was enough safety margin for the men to work in the area. Then, in addition to the usual dosimeters, extra body sensors had been affixed to their outer clothing so later analysis could determine how much radiation they had received and which parts of them had been exposed.

The men stepped on to a special elevatorlike hoist, and were lifted off the ground up along the side of the pile building.

Once in place, they set out to remove the cover from an opening intended for passing uranium cartridges into the interior of the pile. When the hole was open, radiation would stream out. But by taking turns they could place their protected heads in line with the aperture and visually inspect the inside of a working atomic reactor. It was not something the designers had ever intended as a means for determining the extent of a problem, and the true desperation of the situation at Windscale No. 1 was apparent from the willingness of the two specialists to perform the maneuver.

It would be hard to describe the invisible hell which composes the innermost portion of a nuclear reactor core. Given eyes capable of seeing a range of radiation higher up the scale than the human organ can attain, the sight would resemble a surrealist painting. Precise arcs of varied hues would leap up from the cartridges and curve gracefully through the shimmering air to ground themselves at another point. Auras and halos of diverse colors would hover over and encircle the tracks on which the uranium rested. And the very air itself would be alive with tiny fairy motes of flying color.

To a human observer, however, the scene would take on a

127

totally different look. In a dimness broken only by a limited amount of light filtering in from the outside, the rows of capsules, each in their own track, each with identical coloring, would present a dull and listless spectacle. No hint of the restless, seething energy forces would be revealed by visual inspection. The vista is one of another drab, slightly incomprehensible industrial hodgepodge of containers, tracks, gray concrete, and spaced baffling.

No man or woman, however, could stand to take a long look. The damage to his or her system from radiation would be, in a matter of minutes, so intense as to be irreversible. Too long a look at the nuclear Medusa results in certain death.

The men on the lift, though apprehensive, moved with deliberate speed. Cautiously they removed the plug from the wall. Once the opening was clear, Hughes, standing well back from the narrow aperture, raised his eyes in line with the hole and took a long look. What he saw came as a total surprise.

The four cartridges rested securely in their assigned channels. But instead of the dull metallic color they normally had, each was a cherry red. The glow from the surface of the steel container, which encased the now-active uranium, gave off a dim red light. The heat was so intense Tom Hughes took one long look and moved his head out of the line of sight.

He and Gausden talked briefly, digesting the information. Then Gausden decided to inspect the mechanism used to discharge the spent cartridges in the process of normal maintenance. This system refused to function.

Standing a little further back from the opening than Hughes had, Gausden studied the scene for a long minute before withdrawing. His experienced eyes covered the hot, glowing cartridges, the tracks on which they rested, and the ejection mechanism. Finally satisfied, he moved his head and rejoined his companion.

Their conversation was staccato-quick from the emotional stress. According to the visual analysis, the dumping system was jammed because heat had distorted the shape of the outer metal shell of the uranium cartridges, preventing them from sliding down the tracks into the disposal area. Several possible solutions were discussed, and finally the two agreed to attempt to dislodge the jammed units by prodding them with a long metal pole. At this point, they had been on the lift platform for a little less than five minutes.

Six more men, also in special protective clothing, joined them in the narrow space on top of the lift, and working with haste, they rigged a long rod. As one man directed the operation by sight, his eyes in line with the opening, the first dislodgement effort was started—only to be quickly abandoned.

The metal casings were far from fragile in their normal state but, heated to red incandescence, they were in a weakened condition.

The group huddled closely together again, beneath the deadly opening. Another suggestion was agreed upon and they moved into position for a second try. Exposing themselves to varying levels of radioactivity, the men worked through the night in an effort to free the stuck cartridges so they could be removed from the pile.

As they labored, the temperatures in the affected area continued to build. Fearing the fire inside might result in an excess of heat exposure to the uranium cartridges in the adjoining sets of channels, all the capsules were removed from a wide space on both sides, creating an open firebreak. The equipment worked perfectly and the normal cartridges were expelled in accordance with regular operating procedures. The problems in the system were now limited to an area of four channels. But the four distorted cartridges continued to increase in temperature at a slow but dangerously steady rate.

The efforts of the men on the outside lift, who were working directly on the containers, met with a little success. A whole cartridge, albeit one which was showing less tendency to temperature increases, was dislodged and came free into the disposal system. There were now only three left. But these were the hottest and therefore the most severely damaged.

Realizing the situation was becoming very serious, the management of the reactor site met in hurried council.

In addition to several engineers, this quickly assembled group contained Tom Tuohy; Bill Crone, the fire chief; Tom Hughes; and Kenneth Ross. They were faced with a difficult dilemma, and hard decisions had to be made.

First, the temperature in the affected area was still rising. The fire, now contained in one section of the reactor, was showing no signs of spreading but was obviously not going to extinguish itself. There was some likelihood a new release of Wigner energy in another part of the hugh pile might overheat more of the cartridges and they would have a second, or even third hot spot to deal with.

Then there was the matter of the fans. The men working on the hoist lift were exposed to radiation from the inside of the pile. To minimize this and lower the levels of radioactivity in their area, the blowers had been turned on to carry away some of the heat. This added a screen of moving air to the minimal protection of the special clothing, but there was a possibility the fans would add to the atmospheric contamination.

While no one verbalized the full extent of the problem, all present understood if the constant temperature increases were not stopped, the already cherry-red cartridges would become yellow, then white-hot, and finally molten. At this point the out-of-control uranium would be released from its containment and the resulting mess would be impossible to correct. A meltdown situation was slowly taking shape before the weary and astonished eyes of the scientific and management teams.

Action was required. But what they did had to be set off against the damage from additional air releases of radioactive gas and particles. One thing, however, was paramount in everyone's mind: the defective cartridges had to be cooled. And the cooling had to be accomplished in the next twelve hours or the chance of saving what was becoming a very hard situation would be drastically reduced.

After a reasonable amount of debate a number of decisions were reached and an action plan was formulated.

First came protection for the workers in the area, as well as the employees of the Windscale No. 1 pile and the nearby Calder Hall facility. Respiration masks to filter out possibly harmful particles from the air and prevent them from being sucked into their lungs were issued to all staff members. Then, a careful program of employee activity was calculated to enable the various staffs to have adequate warning should massive amounts of contaminants be released.Next, the civilian population, much further away from the site and therefore far less susceptible to exposure, had to be

130

notified on a preliminary basis. The chief constable of Cumberland was alerted to the possibility of an accident. He was asked to arrange for transportation of the factory and construction workers from the site on an emergency basis, if necessary. The constable, not being accustomed to half measures, quickly mobilized a motor pool by commandeering a number of vehicles from the farms in the nearby area. He would prove to be quite ready when called upon.

Additional respirator masks were issued to all workers and visitors at the Calder Hall facility. All personnel not directly engaged in the cooling-off effort were asked to stay indoors and they gathered, finally, in the canteen areas of both Windscale and Calder, to await further information. This preparation was in anticipation of the possible use of the ultimate, and most drastic, solution.

If the now-glowing cartridges could not be dislodged and dropped into safety, or cooled by the use of compressed carbon dioxide—tried earlier on an experimental basis by the management staff and determined to be the next major effort they would make—then the reactor would have to be flooded with water. While there was little or no question the water bath would cool the errant capsules, there was also no doubt this was to be classified as a final, desperate act, to prevent meltdown. It would not be tried until all else failed, because once the water hit the hot containers, a great cloud of steam would be produced. In addition to the risk of a steam explosion which might damage the pile container and cause immense damage to the facility, the hot vapor would rise up the chimney of the Windscale unit and scour millions of radioactive particles from the walls of the flue and other parts of the vent system. The result would be an even more massive discharge of radioactive iodine, strontium, cesium, and other elements into the atmosphere.

These particles would be carried by the prevailing winds and as they settled back to earth, as they eventually must, they would produce contaminated surfaces on grass, tree leaves, other plant life, homes, cars, and farm machinery. Some effort was made at this time to analyze the discharges which had already occurred. The health physics survey van, out since 3:00 P.M. that afternoon running the more quickly performed grab-sample gamma mea-

131

surements, found a very high radioactivity reading at Bailey Bridge, near Sellafield station. Joined at 5:00 P.M. by a second survey vehicle, they managed to take an air sample from the area of Calder Farm road at 11:00 P.M. The reading there was about the same as at the Windscale No. 1 pile site, 23.00 milli-R per hour, or ten times the normally accepted English standard for safe lifetime breathing. The two vans would be called on to continue their sampling work throughout the night and well into the long next day.

Acting on their new plans, the management staff sent for a second supply of compressed carbon dioxide. This gas, used in fire extinguishers, produces a dense, cold cloud which shrouds and smothers a fire. The hope was the cold would be enough to stabilize the hot cartridges and prevent further escalations of temperature.

A crew, working with desperate speed, loaded a truck with cylinders and drove to the Windscale pile. The bulky protective clothing they wore, along with the numerous radiation sensors clipped to their clothes, gave them an otherworldly appearance. It also hindered their movements as they transferred the cylinders to the site where the gas would be released into the hot chamber. Radioactivity levels in the air were high and there was the probability they would go even higher. A number of construction workers who normally were employed outside the buildings had no shelter of their own so were dismissed and evacuated from the area by the constable's commandeered auto brigade.

The air was strained in the main control room as the compressed CO_2 was allowed to flow into the hot area. More and more of the cold gas was delivered into the chamber and for a while it began to appear there was a good possibility of stopping the continued heat buildup. The rate of temperature increase was slowing down—but then it stabilized and fell no further.

Additional carbon dioxide was delivered and forced in, but the result was the same: no real change in the situation.

Finally, after repeated attempts to dislodge the capsules, and the addition of even more carbon dioxide, the management group gave up.

In a midnight meeting they decided to take the ultimate solution: the order was given to flood the compartment. The chief

132

constable was put on alert in case it should become necessary to evacuate the residents of the area.

At 3:44 on the morning of Friday the 11th, the hoses were hooked up. All work through the window into the pile was abandoned and the men who had labored there in unselfish disregard for their personal safety were removed to an examination site where their conditions were evaluated. The facility was teeming with people and the film badge detectors were being developed and processed on the spot to determine the amount of radiation exposure.

A final effort to dislodge the capsules was made before giving the order which would surely release a huge amount of radioactivity. But the steadily increasing temperatures in the most severely affected cartridge indicated further delays would be disastrous.

Friday, at 7:00 A.M. the command was passed to Bill Crone who gave the order to flood the area. Before it was carried out, steps were taken to ensure all workers were under cover. Then a change of shift necessitated another delay. But finally, at about 8:55 A.M., the valves were turned on and everyone shuddered.

The water rushed in, and as it contacted the incandescent steel of the outer surface of the capsules, great clouds of steam hissed into existence. These went billowing up the chimney, only to be followed by more, until the hot portion of the reactor was converting water to steam at a rapid rate. The cooling effects of the flooding were not immediately apparent. So many hot spots had developed that much of the water was vaporized or at least boiling before reaching the glowing cylinders.

Minutes passed. Then quarter-hours. Still the water ran in and still the wayward cylinders gave no evidence of cooling, even though they were now immersed.

Since the men had come down from the elevator platform where the outside wall had been breached, the fans which supplied a limited protection against exposure were shut down. This slowed the drafts through the pile to a minimum. Finally, after an hour, the flooding achieved its aim. First one, then another gauge began to indicate the heating process had been checked.

Tom Tuohy, Ron Gausden, and Ken Ross relaxed for a moment. But a major problem still lay ahead.

133

As predicted, the billowing steam had spewed forth from the chimney, and streams of cloudlike vapor carrying radioactive particles shot skyward.

The trouble came from the breeze. During the final phases of the hookup, before the water was actually flooded in, the winds, which had been blowing out to sea, changed their course and were now forcing the radioactive materials into a north-northwesterly direction at a speed of about 10 knots. In addition, a slight temperature inversion, a not too uncommon occurrence, caused the air higher up from the ground to flow in another direction.

A nuclear rain was about to fall on a large portion of the countryside around the Windscale No. 1 pile. What its effects would be was hard to judge. The theoretical knowledge of this kind of fallout was exactly that—theoretical only. No previous instance was available to use as a basis for predictions.

The flooding of the reactor site continued for a full twenty-four hours, but after the first twelve minutes the level of radioactive materials streaming from the flue was drastically reduced.

The health physics manager now had a multifold problem. A large portion of the countryside in a rough elliptical area centering on the Windscale No. 1 pile had been subjected to a fallout of radioactive particles. This could cause several distinct hazards for humans and animals. First there was the possibility of strong gamma radiation to the whole body. Then there was the likelihood of severe, nonreversible injury due to inhalation of the various hot particles. And finally, there was the risk of ingestion of radioactive materials due to contamination of the food chain. After a careful analysis of the problems involved in the high release levels, and following consultations with other members of the staff, the health physics manager reached the conclusion the materials escaping from the top of the heavily filtered stack were normal fission products, and he therefore planned an approach to handle contamination by iodine and strontium.

By Saturday morning, field workers confirmed this diagnosis. While doing routine samples of the air and radiation levels, one of the two vans made a special check of fresh milk. Since the area around Windscale is heavily dependent upon the production of dairy products, this was a fortuitous and wise move.

The results were stunning.

While the British Government had, at that time, no standards for an acceptable level of radioactive iodine in milk, Dr. Scott-Russell, in a scientific paper, had postulated 0.39 microcurie (μc) per liter as the point at which milk became unsafe for infants.

Early analysis indicated the cows in the affected area had produced milk with traces of Iodine-131, a highly radioactive substance, ranging from just the slightest amount to over 0.48 μc per liter, or 0.09 μc per liter more than the theoretically arrived at maximum safe dosage.

The news got worse. On a subsequent test, several hours later, the figure had gone to an astounding 0.80 μc per liter, more than twice the dose considered harmful.

These results coincided with other information brought to the health physics manager. The air samples taken during the greatest emission, from locations near the spewing stack, indicated a higher level of iodine activity than would normally be present in fission products. The explanation was simple. The elaborate filters located in the bulging top of the flue vent stacks had functioned very well. The particulant matter had to a great extent been trapped. But the iodine, moving in a vapor form, had passed through without much hindrance.

The health physics manager was faced with a real quandary. He could tell, from the iodine content of the milk a process was taking place which was rapidly contaminating a basic food substance. As the iodine settled to earth, it coated the blades of grass in the pastures around the site. The cows ate the grass and as they digested it, absorbed the iodine into their systems, where it in turn was passed into the milk they gave. The more grass they ate, the stronger the concentrations became. In short, the cows were giving radioactive milk and the radioactivity was in a form which would be readily absorbed and stored in the body of all humans who consumed it.

But there was no standard for defining an "acceptable lifetime level" for radioactive iodine in milk. To halt the production of the many dairies in the area would have a dire influence on the economy of the farmers and the region, but it appeared to be the sole choice of action. Not only was the milk contaminated, but cheese and other products made from the milk, which had an even wider distribution, would be heavily laced with the radioactive iodine.

135

The decision to cut into the livelihood of an entire region was far too big to be made by a single man. But the need for action was clear and—to his everlasting credit—the health physics manager took it.

His interview with Tom Tuohy, the works general manager, was brief but full of impact. Upon explaining his fears and after asking a few questions, the works manager reacted with decisive swiftness.

A group of medical experts with knowledge of radiation effects on humans was gathered together. Several of the top men could not be present so their opinions were sought by telephone. They were charged with the establishment of a limit for the presence of radioactive iodine in milk. This step was necessary if all milk produced in the area was to be pulled from circulation. Such a dramatic act was bound to provoke the question "Who says so?" and an authoritative answer had to be ready. The specialists spent several difficult hours forging a standard. While many views were expressed, they finally managed to agree a level of $0.1\,\mu c$ per liter of milk was the absolute maximum allowable. This was a content point which had been surpassed by 800 percent on Saturday, the second day of field testing.

The figure, while considered low by some experts, came from an analysis of the probable absorption into the thyroid glands of young children. Given the fact that iodine has a strong tendency to be stored in this organ, and taking into account the facility of the thyroid to build up iodine over a long period of time, all felt that any great concentrations should be avoided. These findings were quickly delivered to a higher echelon, and members of the Atomic Energy Authority and the Medical Research Council agreed with the limitation.

While the scientific work was in progress, another series of meetings, also arranged by Tuohy, was held to discuss the political and social implications of a ban.

At this point, even though the specially equipped vans were out sampling additional milk supplies, no exact knowledge of the range of the radioactive fallout was available. For immediate purposes a circle about two miles in radius was drawn and this was the area under debate. A total of twelve milk producers operated in the zone and they would bear the brunt of at least the first proscriptions.

The field testing took time. And still more time was required to study the data. But by shortcutting unnecessary debate the scientists, engineers, technicians and politicians were able to combine their findings, so by 11:00 P.M. on Saturday, October 12, they had reached a three-part decision. All milk from the dozen suppliers in the two-mile field would be declared contaminated and disposed of. The sampling vans would work outward from this immediate area and define the furthermost limits of the fallout, basing their judgment on the iodine levels of fresh milk. If they exceeded the 0.1 μc-per-liter critical level, these supplies would also be declared unfit for human consumption and confiscated. Finally, additional support would be called in to take random samples in a huge geographic area as far away as the coasts of Lancashire, North Wales, the Isle of Man, and into Yorkshire and the southernmost portions of Scotland.

At this same time, additional studies of the iodine content of such other farm produce as eggs, vegetables, and some meats were authorized, although these were given a lower priority.

Shortly after the group agreement, action was taken. John Bateman, the dairy farmer, was awakened at 1:30 Sunday morning by a commotion in his front yard. Motorcycle policemen knocked on his door and when he answered, told him, with measured politeness, about the contamination problem. The Cumberland police, assisting the Milk Marketing Board, moved in and seized the milk at twelve farms. The individual dairymen were stunned though cooperative. They had no wish to place a harmful product on the market, but they were noticeably concerned about their livelihood. A portion of their loss would be protected by the British Government, but the long-term damage, which would affect both the consumer's attitude toward their milk and other livestockmen's feelings toward their herd animals, was a problem which would trouble them for years.

As the testing vans expanded their radius of operations the area of known contamination began to grow, and in stage after successive stage the original two-mile radius became three, then five, and finally the idea of a circular ring of fallout gave way to a clearer picture. The pattern became one of a long ellipse with the Windscale No. 1 pile at the upper end. It stretched to cover a parcel of land thirty miles long, ten miles wide to the south, and six miles broad in its northern extreme. The upper tip of the

137

area terminated about six miles from Windscale No.1, and the southernmost point included the Barrow Peninsula. The total was over two hundred square miles of fine farm and dairy lands placed under quarantine. All milk shipments were halted. Months later, while the radioactive levels in most areas had fallen, the milk ban still remained in effect throughout the fallout zone.

Other testing activity was under way at this same time.

The workers who had been on site during the various periods in which the radioactive materials were spewing into the air were being given close study to determine if they had been exposed to the fallout. While several showed signs of hair and hand contamination, they all responded to standard treatments, which consisted of degreasing and scrubbing the skin. Large doses of a heavy iodine drug were administered to prevent the thyroid from absorbing radioactive iodine from the air.

Tests were also run on residents in the areas near the facility to discover the extent of their exposure. Several cyclists going to work along the track near Seascale on Friday morning had fairly high counts on their outer clothes but the levels were well into the so-called "safe" zones. Other individuals also had some low-level contamination, but no problems were encountered and all were able to clean up to normal standards.

An ongoing radiation test for iodine in the thyroid was instituted for all men, women, and children in the exposed area, and when one child showed a 0.28 μc level, additional tests were planned. A special study by the Medical Research Council was also begun, and this work continued into 1977.

A great national uproar met the news of the accident. What had been so long feared was now a fact. And tempers, as well as emotions, ran high.

Special government studies and committees of examination were formed and charged with the responsibility of discovering every detail of the incident, as well as projecting the probable damages to the economy.

These investigations were moving at a normal governmental pace when the real blockbuster was dropped.

Danish scientists, concerned by the news of the Windscale accident, studied the pattern of air currents in the upper atmosphere and began checking milk produced on the part of their peninsula closest to the disaster site.

138

By October 23, their results threatened to bring about an international crisis.

The tests revealed levels of contamination equal to or greater than those which had been reported in the British press. The affected area, while smaller than the 200 plus square miles under ban in England, was of a considerable size—especially since it was centered on one of their country's most vital milk and dairy areas.

Increases in both Strontium-90 and other radioactive trace materials had been found in milk before and the causes had been attributed to the test explosions of nuclear weapons. The around-the-world travel of the hot debris cast upward into the atmosphere in the famous "mushroom"-shaped cloud had been carefully tracked on a number of occasions, and studies had shown wide belts across the face of the earth were contaminated for short periods of time. But the Windscale pollution was another matter entirely. In this instance the contamination rates were of sufficient magnitude to render the milk unfit for human consumption. And there was no guess as to how long the levels would remain intolerably high.

New questions of international law were raised by the incident. Was the government of one country responsible for the losses of either private citizens or the government of another nation due to the accidental release of radioactive substances? Or was the release merely an act of God? These questions, perhaps regrettably, were never answered. Discussions took place in Parliament and an official government policy formulated which acknowledged only minimal responsibility for the accident in the first place and no acceptance of liability in the second. Likewise, in Denmark, governmental councils convened and although incensed by the incident, only the most civil and noncontroversial communiqué was drafted and sent. Judging from the newspaper coverage of the time, the matter came up and then, in a matter of days, after a violent series of threats, complaints, and intense media coverage drifted into oblivion.

Meanwhile work continued on the reactor. New designs were proposed for a second plutonium production unit on the same site—with several noticeable improvements—showing the government's commitment to the area remained unchanged.

More than 600 dairy farmers were affected by the ban, which

139

continued over the entire 200-mile area for longer than nine months, and over a part of the contamination zone nearest the Windscale site for more than a year. During the time the total area was under quarantine, an estimated $11,000 U.S. dollars' worth of milk was daily poured down drains and allowed to flow into the sea. The financial loss exceeded $3 million from this one product alone. And those were 1957 dollars.

Even further financial devastation awaited the farmers and landowners near the center of the contamination. Breed cattle from these farms fell drastically in value, as did the selling prices of the land itself. And financial lending institutions, which had long considered the region prime for investment purposes, began to look with disfavor on further loans for capital construction and expansion.

Fortunately, given sufficient time, things do return to normal. But years were required to re-establish confidence in the areas of maximum contamination.

Some final questions about the Windscale disaster remain.

Did the reactor protection systems work as they should? Was there a real chance of an uncontrolled chain reaction which would progress to disaster proportions? And did the engineers, using the incident as a stimulus, improve their technology?

In answer to the last question, the process upon which the Windscale reactor was based is no longer in use. Considered to be too intrinsically dangerous, air cooling has been largely replaced by other systems. Scientists and engineers knew long in advance of the Windscale incident that the graphite pile was destined to give major troubles. And a great part of the informed scientific community did not view the Windscale disaster with surprise. While saddened by the accident, the majority of these knowledgeable people were relieved the consequences had not been even worse.

It is apparent many of the components of the system were either inadequate or nonoperational. And it is equally clear, from the number of instances in which workers were exposed to potentially intolerable levels of radiation, the management of the pile did foresee a developing disaster. That they were finally able to contain the problem by direct human exposure and emergency measures developed on the spot says a great deal about their ability to think calmly in periods of duress, but little for the thought which went into the design in the first place. The attempt to settle the situation by actually breaking through the retaining wall into a radioactive no-man's-land, and then manually cooling the wayward cartridges by directing streams of water onto them, was both a brave and a last-ditch solution to a difficulty which was rapidly assuming the proportions of an ultimate accident. Allowed to continue, the heat buildup would finally have caused the release of an almost uncalculable amount of radioactive "ash." Uranium oxide is highly toxic. At the levels it was being produced by the Windscale overheating it wasn't too serious a matter. Most of the particulants were trapped by the elaborate filters in the tall, bulging ventilating stacks. Further progression of the problem, however, would have resulted in a far different ending. Finally reaching their overload points, the filters would have then started to bypass an ever-increasing amount of waste, and this virulent material would have been forced into the atmosphere in quantity. London, 300 miles to the south, recorded 20-fold increases in radioactivity during the incident, and might have become too contaminated to allow for the continued residence of humans there.

Two things have to be taken into consideration when examining this scare story. First, cooling was attained, albeit by unusual means, under emergency conditions. And second, great technological strides have been made in the ensuing years. The designs of both the basic reactors and the safety equipment have improved immeasurably.

But Windscale again shows us it is just possible we may not, at any given time in our technological development, be as advanced as we think we are.

Had the press questioned the safety of the Windscale No. 1 pile before the incident, and asked specifically about the Wigner release process, the answer would have been totally affirmative, based on the management's honest belief every contingency had been analyzed, discussed, and dealt with. It seems to be the nature of technicians, no matter how well-intentioned they are, to take somewhat simplistic attitudes toward the various processes they operate. Their very familiarity with the technology produces a feeling of safe acceptance. Suggestions of potential

141

danger or problems generally tend to be rebuffed by an increase in references to the "fail-safe" nature of the designs, and finally, an irritated "This thing is safer than a loaf of bread" kind of remark.

In other words, at any point in our development of nuclear knowledge the processes in use represent a technology which is at one and the same time advanced but untried. And since safety has always been a basic criterion for design in this field, any suggestion the system might not be sufficiently safe is bound to come up against an especially positive assertion that there is absolutely no way the device might malfunction or fail.

But such a possibility, no matter how remote, is always present. And the designers, no matter how positive, know it.

10

■ Limestone County is a pretty place, centered in the state of Alabama along its northernmost border with Tennessee. The land is hilly and runs down to the backed-up waters of the Tennessee River at its southern end, where it joins into Morgan County, named after an early settler.

A lot of the names in the area made headlines in the 1930s when the Tennessee Valley Authority (TVA) began to dam up the mighty river in a multistate project designed to stop the disastrous annual flooding and produce electric power for thousands of homes and new industries.

The countryside is partly open, partly wooded, and was peopled by farmers who grew cotton and corn and raised cattle. In the war years of the 1940s, it changed a little as industry moved into Decatur and a few other small towns, bringing jobs in man-ufacturing plants powered by the cheap electricity of the TVA. But in the 1970s, things changed a lot more.

Just south of where the Elk River joins the dam-controlled waterway, close to the red clay banks of the Tennessee, a new kind of power plant was erected. The staff of this one have titles like "Reactor Operator," "Nuclear Engineer," and "Environs Direc-tor." They talk in a jargon which includes words and phrases like "scram," "ECCS," and "core melt."

The TVA, long a leader in the production of electricity through hydroelectric power, decided in the late 1950s to expand and maintain its pre-eminent position by designing and building the world's biggest nuclear power station. It would be ten times larger than any plant planned for the 1960s. The total facility would house three giant reactors, each capable of producing about 1,100 megawatts of power to serve two million people in the area with cheap, clean electricity.

The group that dammed the mighty Tennessee had little

trouble using their Washington connections to procure additional federal funds, and before long their second major dream in 40 years became a reality.

And Browns Ferry became an historic site.

The TVA, a branch of the federal government founded by Franklin Roosevelt in 1933, had for years cooperated with the AEC and the Defense Department in the development of nuclear facilities within its boundaries. Oak Ridge, Tennessee, is far to the north of Limestone County, and houses one of the greatest concentrations of atomic research and nuclear materials production facilities in the entire world. It must have seemed a logical step to turn to atom splitting when the demand for even cheaper electricity threatened the future growth of the Authority's protected area.

The Browns Ferry facility was celebrated at its opening in August 1974 as one of the best-constructed, state-of-the-art, safety-oriented installations of its kind. The rhetoric seemed to please a lot of people, made good press, and was right, as far as it went.

But the facility, like many other operating plants, contained within its massive reinforced concrete structures a basic design flaw.

Picture a room 180 feet long, 35 feet wide, and 11 feet high. This cement-walled tunnel stretches away into the distance, cutting through the innards of the plant and passing directly below the main control room. Its purpose is simple. It is economically more sound, from the standpoints both of construction and maintenance, to build a single open space which runs through the entire complex than to lay in separate multitudes of conduit to carry the thousands of miles of wiring needed to control the valves, pumps, rods, and vital instrumentation required to operate the complicated reactor, pile, and generating equipment.

The long, open room, brightly lighted from above, is a maze. Along its dun-colored walls, and crisscrossing at random from side to side down the entire length, are shallow metal troughs made of gray galvanized iron. In these flat trays rest thousands of multi-colored cables, connected at one end to a valve or gauge and at the other to a control or readout in the main operating center. It's impossible, even with the 11-foot-high ceiling, for a man to walk erect through the orderly jumble of steel, plastic, and

144

wire. Electricians labored in the area, known as the cable spreader room, for more than a year, installing each of the lines required by the construction blueprints into a Chinese puzzle of infinite complexity. Each wire is numbered, carefully identified, properly bundled with its brothers, starts where it should start, and ends where the plans dictate. One look at the accumulation quickly banishes any thought the business of generating electricity from atomic power is a simple process.

But the mass of wiring is relatively easy to comprehend. It is logical, and any licensed electrician could take the proper drawings and make order from the seeming chaos. People work in the bright room year around, installing new items or modifying existing hookups to improve performance. They are so busy, in fact, a shift is often on duty on Saturday attacking one project or another. Their comings and goings are natural enough to be unnoticed and a new man rapidly acclimates himself to working in the confinement of the huge space.

In a sense, the men exist in a world of their own. Fully unionized, they hold periodic meetings with superiors to discuss the proper ways to perform their complex tasks, but they work pretty much at their own speed, with breaks for coffee and a main meal. They also devise their own techniques for achieving their goals—techniques which may or may not be directly approved by the inspection and management teams who attend the meetings but seldom enter the confusion of the main spreader room.

The Nuclear Regulatory Commission, assigned to monitor the safety preparations and operation of the facility, only occasionally sends anyone into the work areas of the plant. Secure in their offices, its members peruse papers showing the required number of fire drills have been performed, or analyze reports which indicate the number of days the facility has operated without mishap or worker injury.

Many employees of the NRC would be of little use in the actual functioning area of the plant anyway, as only a limited percentage are technically trained or specifically oriented to the components of nuclear power production.

As in most industries, the shift workmen and their managers are left to their own resources in operating the plant. To varying degrees these people come to know and respect the equipment they are assigned to run. They develop a possessive feel for the facility and its operation. The controls take on seemingly human traits and the idiosyncrasy of each meter, gauge, switch, or computer-operated event sequence is well known and thoroughly discussed. The plant becomes as familiar to the men as their own homes. They know its every nook and cranny and are able to walk from one place to another in the wide-spread complex without a guide or even conscious thought. This familiarity breeds a relaxed air. Even the tension each man may have felt in the early stages of dealing with not one but three huge atomic piles fades with repetition after repetition of control sequences and a never-ending string of successful uneventful operations.

From the lowest member of a cleanup crew to the plant operating superintendent, the job has its routine aspects. And the Browns Ferry installation is so huge no single person can possibly monitor the daily happenings. Men come and go on their duties and assignments. Shifts change with only short briefings, and then a new team is manning the unit. Construction work goes on constantly as the third reactor comes on stream. New office space is needed as the staff grows. And the number of electrical connections and improvements in the wiring system continually increases.

Everything went along fine until March 22, 1975.

It all started simply enough. Months before, when construction began on reactor number three, an airtight partition had been installed between Unit 2 and the new building to prevent any leakage of possibly contaminated air back into one of the working control rooms.

All air movement in the plant is designed to flow toward the areas of highest radioactivity. This protects the workers, and, in case of a leak, prevents contamination from spreading outward, as the air pressure helps pull it in.

The time had come to remove this partition between the two units, but before authorization could be granted, there had to be a series of leakage tests to find passageways for air between the buildings. The Division of Power Production (DPP) was charged with this job. The tests quickly indicated there were several areas where leaks had occurred, and these would have to be corrected before the airtight partition was pulled down.

The Division of Engineering Construction, informed of the problem, issued a work plan, number 2,892, requiring all leaks be

146

identified, listed, sealed off, and the work verified by an engineer.

It seemed like such a simple order—until the electricians came to the main cable spreader room.

The maze of troughs, along with the shadows they threw and the darkness in the corners away from the ceiling lights, made for slow going. Many normal leak-testing devices were unusable because the men simply could not see to read them. Smoke was tried, along with soapy water which would show bubbles. But the most favored technique utilized a small candle held near a suspected airflow. The flame, agitated by the passage of air, would flicker and follow the draft, indicating the precise location of a problem.

Hard as it was, working in the tight confinement between the stacks of trays, the men were successful in developing a primary list of leak points. Work started on patching these, and as one by one they were secured, the air pressure in the room increased by small but measurable amounts. This, in turn, caused smaller points, which had gone unnoticed or had not leaked before, to show up on repeated inspection.

The job was a long one but its completion was vital to the start-up date for the third reactor. They had to achieve a maximum state of airtightness before the fueling-up process began.

Management was concerned both with the magnitude of the problem and the time it was taking to solve it. To hurry the process along, they assigned several engineering aides, or inspectors, to the job of working directly with the electricians testing for leaks and signing them off as they were sealed.

Many of the leaking places were caused by the addition of new wiring. When it came time to install a needed line, the electricians made a hole through the sealant foam, at the point where the loom of wires going to the same place passed through the wall, and simply pushed in a new cable. Little attention was paid to sealing up these newly punched passages as things were still in a construction phase.

■ March 22 was a balmy Alabama Saturday. Scattered clouds spotted the deep blue of the sky and visibility was clear across the rolling, pollution-free landscape. It was a great day to be off work and outdoors.

147

The six men working in the artificial light of the cable spreading room could see nothing but the miles of wiring lying in the gray trays. All thoughts of recreation had vanished from their minds as they worked on what was normally an off day, trying to seal a seemingly endless number of leaks. Dealing with some of the harder-to-locate problem areas, the teams, working in twos, were checking places where cables and special conduits passed through the thick concrete walls of the room on their way to and from their connection points.

An engineering aide named Larry Hargett and an electrician were at a place where ten cable trays, placed in two vertical rows of five each, entered the Unit 1 reactor building. Hargett, a twenty-year-old who had been on the job for only two days, was working in a dim light, using a candle to detect the airflow.

The flame flickered, and from time to time his eyes, strained by watching the erratic brightness, would blink. He had taken the candle from the electrician because he was in a better position to reach a suspect place.

Suddenly, the flame darted off to one side where several cables had been punched through after the original fireproofing had been installed. Intrigued, the aide leaned farther forward to bringing the candle closer. The flame was pulled horizontal and Hargett knew he had found a relatively big hole.

Removing the candle, he checked to see if the electrician had noted the spot, then moved out of the way as far as he could in his confined placement above the floor to allow the second man access to the leak. The electrician, stretching, could not reach the penetration point because it was deeply recessed into the solid wall.

Seeing the other man' s dilemma, Hargett asked if he could help. Nodding, the electrician tore off a couple of hunks of the two-inch-thick sheet of polyurethane foam he carried and passed them up. The aide, working rapidly, stuffed the pliant bulk into the hole, then brought up his candle to check the result.

Moving the flame to within about an inch of the protruding material, he watched. Then, to his surprise, the fire, following the airflow, was pulled into the leak. The foam sizzled and before Hargett could remove the candle, the material burst into flame. He was not overly alarmed. Small fires had been started in the spreader room before, during similar operations, and had always

148

been extinguished. Many of the men had pinched them out with their fingers. This fire was very small. Sputtering and dropping little flaming streamers, it flickered brightly.

Calling down to the waiting electrician, who could see something had happened, Hargett told him there was a small fire. The man below handed up a flashlight and Hargett, using the blunt handle end, tried to crush out the flames. After a minute he gave up. The area was too deeply recessed to reach easily, and every time he squelched one burning area another cropped up.

Nearby, a third man heard the commotion and quickly passed over a handful of rags. The wall where the fire burned was about thirty inches thick and the inspector stuffed the rags into the four-by-five-inch opening to try and smother the small blaze. After a moment's wait he removed the wadding. Peering into the narrow aperture, he could see the flame still flickering. The rags in his hand were starting to smolder from the intense heat. He dropped them quickly to the electrician below, who stamped on them as he called for a fire extinguisher.

The incident had lasted only 90 seconds. Someone quickly brought the electrician a carbon dioxide unit which he hoisted up to the waiting Hargett, who forced the hornlike nozzle into the opening and emptied the device in a single, loud, sustained blast. He passed the white-frosted canister down to the waiting men and again peered into the now-hot opening.

Inside, he could see no sign of flames, and relaxed. After a few seconds, he looked in again to reassure himself everything was extinguished. To his surprise, he saw more smoke, followed by the first fingers of a small flame.

The fire had heated the copper wiring to the point at which it could ignite its own insulation. Carbon dioxide had forced out all the oxygen in the small hole but it had been sucked downward by the constant draft. As soon as air returned, the blaze re-ignited.

From his position Hargett could see the fire had managed to spread outward from the hole through the wall into the reactor room itself. He shouted down to his companions and two crewmen hurriedly left the cable spreading room to fight the fire in the reactor building on the other side of the wall.

One, seeking additional fire-fighting equipment, stopped at Security Post 8D. Running up, he grabbed the available extinguisher and began to carry it away.

149

The guard, seeing a problem and conducting a hurried, excited conversation with the man, stepped inside his station and dialed the telephone number which was posted on the plant's emergency procedures card. Nothing happened. It later proved to be a wrong number. Confused, the guard then dialed the shift engineer's office to report the incident. The man on the other end of the phone punched in 299, the correct fire-alarm code, and handed the telephone to the shift engineer (SE) so he could ring the reactor operator.

Luckily, at this point the shift engineer's office was on the "inside the plant PAX" hookup. Construction was still in progress and a totally separate telephone system, connecting the work crews, was also in operation.

There was now increased action in the spreader room. By this time, Hargett's electrician partner had brought him two additional fire extinguishers. Each was exhausted and then tossed aside. Volumes of acrid smoke were coming from the small hole, and when the electrician passed up a third unit, the inspector could hear the hissing roar of other fire extinguishers being released on the far side of the wall in the reactor building. He fired off his bottle, and sending it down, took up a fourth. It, too, was shot into the hole, with no more effect.

In a moment of silence after the loud sound caused by the release of the fire-fighting chemical, the inspector could hear an ominous crackle. The burning insulation, sputtering and throwing off tiny drops of pure flame which left smoke trails in the air, was out of control.

Hearing the alarm, the shift engineer hastily moved to a control box which would release the built-in Cardox fire-extinguishing material. He was astounded to find the box still had a metal plate over its glass window, installed to prevent accidental breakage during construction. It took some time to remove the protective cover with a screwdriver.

Hargett, still near the origin of the blaze, heard the fire alarm sound, indicating someone was going to flood the spreading room with CO_2 gas. He and the others quickly evacuated the area to avoid being caught in it.

The assistant shift engineer (ASE), after determining there were no men still inside the spreader room, attempted to activate the built-in system. But it refused to function. The control box had, for some reason, been un-wired. Thinking quickly, the ASE ran to the east door of the room, where a second box was located. He went through the same routine, and this time was rewarded by a whoosh of carbon dioxide.

It was now just after 12:40 P.M. The fire had been burning for 20 minutes.

The two men who had left the cable spreader area were joined by a third worker on their way to the reactor room. They arrived carrying an extinguisher each. The smoky fire was instantly located in a set of wiring trays about 20 feet above the floor. A ladder was nearby and one of the crew moved it into position. A second man quickly climbed up and emptied a dry chemical extinguisher onto the flames. The fire was knocked down but quickly rekindled and was soon flickering again. Burning insulation was giving off a greasy, black smoke.

The third worker alerted other people in the reactor building to the problem and returned to find the man who had been on the ladder forced down by the noxious fumes.

An assistant shift engineer arrived, and aided by another man, released both carbon dioxide and dry chemical extinguishers onto the now rapidly spreading fire. The smoke and fumes were intolerable, so the ASE returned to the floor. Taking charge of the action, he sent for special backpack breathing apparatus, and for the next five minutes he and the others tried to extinguish the blaze from floor level.

Once the breathing devices arrived the men donned the masks and were able to deal more directly with the problem. But the flames had gained a strong foothold. More and far denser smoke billowed from the cable trays, obscuring vision until there was no way to maintain a further effort.

The men, thrust away from the base of the blaze by a constant outpouring of fumes from the burning plastic insulation, finally reached the end of their endurance. They could no longer approach the source and were forced down the room to a stand near some large heat exchangers.

At 12:35 P.M., when the fire alarm sounded in the plant control room, there was no hint this shift would be anything out of the ordinary. Since the station ran on a twenty-four-hour, seven-day-a-week basis, the operators of the two reactors worked staggered hours. A team of 17 experts manned their stations on nights, Sundays, holidays, and the normal work week.

News of the fire caused no panic among the technicians in the large futuristically designed, well-lighted room. One assistant shift engineer turned on the manual alarm so it would ring continuously throughout the complex, alerting other personnel. And another ASE began making announcements on the internal loudspeaker system calling for the individuals who were assigned to fire suppression to go to their stations. Back in the shift engineer's office the time of the alarm was noted in the logbook.

Due to the way the incident was reported, no one in the control area, directly above the main cable spreader room and therefore right above the fire, had any idea where the blaze was—or how serious it might be. An assistant shift engineer was unhurriedly given the assignment of locating the problem and reporting back.

No consideration was given to shutting down the reactors, as there did not seem to be any difficulty directly related to the control system.

Two million people were relying on the power generated by the facility and more than a hint of trouble is required to cause the operators to close down the units and deny the populace the energy it desires.

The main plant control at Browns Ferry is designed to act as a point from which the equipment required for all three reactors can be operated. Looking somewhat like a TV program's idea of a spaceship's command bridge, the room is a complex of light displays and buttons.

The operators can not only command various valves, pumps, and other electromechanical devices to start and stop, they can regulate the speed at which they perform and also monitor vital pressures and temperatures as well. Since both reactor one and reactor two were on line and operational, two complete teams occupied the 180-foot-long room.

To the uninitiated, the place seems to be a modern confusion of colorful panels, dials, gauges, and switches. Each individual system has its primary and redundant secondary controls. The men on duty are more than mere needle-watchers. They have

152

been schooled in the complexities of the system, can make mental estimations of the probable causes of malfunctions, and are versed in the steps to correct problems as they arise.

The assistant shift engineer sent to look for the location of the fire did not have far to go. As soon as he reached the lower floor, he saw several running men and went to investigate. Minutes later, he called the shift engineer's office with the news the fire was right below the control room and had spread into the reactor building itself.

The report was disconcerting but no one panicked. After all, they were contained inside a massive concrete structure and unless the fire was very bad, there was no real danger.

Then the first of what was to become a series of unusual incidents occurred.

An alarm went off. Its strident sound carried above the normal noise of the center and the Unit 1 reactor operator moved to investigate. Before he could react to the first problem, a second, and then a third alarm went off.

He studied the situation. The emergency core cooling system (ECCS) had been triggered. His eyes swept the indicators. A digital clock on the panel showed 12:40 P.M. The water level covering the reactor registered normal. This was one of the most crucial points. As long as the depth of water completely covered the top of the pile, the ultimate danger could be averted. The steam pressures looked normal too. He checked to see if someone had mistakenly turned on the standby equipment, but there was no sign of error.

Two men, staring at the control panels, began to discuss the possibility of "scramming" the reactor, their term for an emergency shutdown brought about by inserting the control rods to stop the radioactive energy cycle. The reactor operator was scanning his instruments when a new alarm rang. He checked the source, and found another portion of the emergency core cooling system had started of its own volition. Since pressures still seemed normal in the core, he attempted to shut the units down, but as soon as he would release the controls, they would restart.

There was some confusion in the room when, to everyone's surprise, smoke started rolling out from under control panel 9-3, which contained the guts for the emergency core cooling. Green and red lights on various boards began to act erratically. They

153

would shine brightly, apparently at random, then dim, flicker, and glow intensely again. None of the emergency equipment would stay off. Almost as if the units had minds of their own, they would stop, pause a moment, then re-start, causing their indicator lights to glow and add to the chaos.

The 17-man crew was on full alert even though only reactor control Unit 1 was showing any malfunction. The second unit seemed to be operating in a routine fashion.

At 12:48 P.M., an ASE noticed the power produced by number one had fallen from 1,100 to 700 millivolts. The smoke from under panel 9-3 was starting to fill the room with an acrid stench caused by the burning insulation. The next surprise came seconds later. Entire electrical boards began to flicker, then one at a time, go blank. As indicated by their appearance, they were dead. With the fire burning the wiring, the room was losing its means of controlling the pile.

Another gauge showed the level of water over the core to be two to three inches above normal, so the operator took corrective procedures.

The action began to accelerate. Men were moving swiftly in the background, some telephoning for additional aid, others holding quick conferences. Voices showed tension.

One by one, a number of relief valves necessary to regulate pressure in the core were lost. Fully one-half of the reactor protective systems were out of order and of no use.

Alarms rang all along the control area and the smoke was reaching a point where it had become more than a nuisance—it was now making breathing hard for the men in the long room.

One of the hurried consultations broke up and the shift engineer moved alongside the operator of Unit 1. He spoke in a loud voice to be heard above the clamor:

"Let's scram the unit."

Several men heard his shout and moved to their stations. With all the noise and the growing confusion, it took a moment, then, with deft motions, the operator manually performed the scram operation, slamming home the control rods. He threw the reactor mode switch to its shutdown position and turned to nod to the shift engineer. But the SE was already on the telephone, reporting the conditions in the plant to various supervisors.

At 12:53 P.M., positive signals were received. The control

154

rods were confirmed to be fully inserted. Reactor Unit 1 was essentially down. But there were still very serious problems. Even though the main energy source was shut off, residual power would remain at very high temperatures for a long time to come. Cooling had to be maintained. The 200 inches of water on top of the reactor pile, and the pressure levels in the reactor vessel, had to stay at a semi-operational level or the core, allowed to go its own way without cooling, would turn into a 5,000-degree puddle of molten uranium.

Everyone in the control room relaxed when the unit operator announced the successful scramming of his pile. But a look at the various control boards showed a growing problem. The normal water system which fed coolant to the core was out. The high-pressure emergency cooling system was out. The reactor core spray was out. The low-pressure emergency core cooling system—out. The core reactor isolation cooling system—out. Worse, most of the various instruments indicating conditions inside the unit were also lost. In short, every one of the normal ways to feed water into and take water out of the reactor to maintain its cooling were nonfunctional.

But other crossover arrangements could be improvised. A single high-pressure pump, used to drive the control rods, was switched to deliver its flow into the crucial area. No one could estimate if it would be effective. The volume it could produce was much less than normal, but it was the best quick solution.

During this sequence of malfunctions people were coming and going between the control room and the fire. One assistant shift engineer returned to the scene of the blaze in the reactor building to take charge of the crew now trying to extinguish the smoldering wiring.

The smoke in the control room was becoming intolerable. Each time the crew below in the main spreader room would activate the Cardox system to kill the flames, the extra air pressure from the carbon dioxide would force more smoke into the main control center. There was a serious question whether the men could stay there and remain on duty.

At 12:55 P.M., just half an hour after the start of the fire, all ability to monitor the radioactive happenings in the pile was irrevocably lost. Reactor number one was operating blind. And Unit 2 was starting to give trouble as well. The same strange flickering of panel lights had started. Worse, alarms indicating something had gone wrong in the wiring system of the diesel-powered generators sounded, quieted, then came on again. The Unit 2 operator called the shift engineer. He didn't think the engines would start.

The shift engineer, meanwhile, had been frantically busy. In addition to directing a call for assistance to a number of off-duty personnel, he had been asking for advice from his supervisors on how to deal with the growing emergency. The thickening smoke had reduced visibility in the long room and several of the staff were violently coughing.

The Unit 2 operator was faced with an increasing problem. Reactor power was decreasing at a phenomenal rate and a number of alarms were calling for immediate attention. Most of the indicating lights had failed and the few still on were starting to flicker. After a hurried call to the shift engineer, who had to round up several other technicians, the Unit 2 operator gave orders to scram that pile and started his own shutdown procedure.

The fire-fighting teams were being beaten back and the situation was nowhere near in hand. It was hard to tell how much worse the matter might become, so all personnel not directly concerned with the operation of the reactor units were evacuated. A careful head count was taken at 1:15 P.M., and it was ascertained all were accounted for.

Breathing apparatus was rushed to the control room, and even though the units needed frequent changes because they would run low on air, they gave the operators a chance to stay in the area. The foul stench of burning cable insulation had seeped everywhere and it was now hard to see through the haze.

The word had been passed down into the main cable spreader room about the smoke in the control center, so the fire fighters had stopped using the Cardox system. This helped, but the air was still only barely breathable.

More and more electrical boards on the Unit 1 side were going out. Then another problem began to surface. The pressures in the reactor chamber were building higher and higher. In a matter of minutes they had increased to well over 1,000 pounds per square inch (psi), and only the control drive rod pump had the capacity to force water in against this immense force.

Trial after trial to re-rig power to crucial valves failed, so the

156

decision was made to try and operate them manually by sending out some of the now-arriving off-duty technicians and maintaining a linkup with them by telephone.

Several workers were at their stations when all electrical power for the reactor building failed, and the elevator stopped.

One operator successfully opened the proper sequence of main steam valves by hand. Pressure inside the unit dropped to about 850 psi.

But the quantity of water being delivered by the single pump was insufficient to maintain the water level on top of the core. This was slowly but steadily falling.

Unit 2 was having more problems. Successfully shut down, it too was experiencing the failure of its emergency core cooling system. Drastic steps were being taken to stay ahead of the trouble.

The operator of Unit 1 and the shift engineer held another hurried conference. Both were highly concerned over the loss of water from the core. They started to work out additional possible linkups which could handle the problem, but it was a difficult balance. The heat in the reactor would turn water to steam at normal atmospheric pressures, so they had to maintain a high enough pressure inside the unit to keep the water in a fluid state but lower it to a level where some other pumps could be used to force more liquid into the core.

A new emergency arose. Part of the reactor unit, called the torus, was overheating from the failures and unusually high pressures. It was mandatory this be cooled as rapidly as possible to prevent a steam explosion large enough to damage the main containment structure and release huge volumes of radiation into the environment, but the ways available were by now severely limited.

Working in a logical sequence under growing duress, the engineers on duty determined the pumps normally used to scavenge the water formed as condensation would be more than adequate to hold the level above the core at the required point, so they hooked up a relay to bring these units into play. But by this time another problem had developed.

The temporary drop in pressure in the main reactor vessel reversed itself and the meters climbed steadily to 1,080 psi, then seemed to stabilize at about 1,100 psi.

This was far too much for the condensate pumps to overcome. Designed to function at about 350 psi, these units simply could not push water in against the growing force in the main reactor chamber.

This same pressure also made the problem with the torus more serious. Water was being pressed out by the enormous pressures in the reactor above and it was growing hotter by the moment. A failure would result in a completely hopeless situation. The pile would be unprotected and a meltdown would be imminent.

The control rod drive pump, operating at full power, offered the only hope. Normally, there was more than one unit in this series to provide a backup, but the controls were effective only on the primary.

At about this point the process computer, which notes each activity instigated from the control room, failed. No further records of the action were made until about 4:00 P.M. The loss was not significant to the men on duty but it meant all reconstruction of the event would have to be accomplished by memory, later.

Smoke in the control area had lessened but the stench was still bad. The technical staff pinpointed the problem of pressure inside the reactor as the most serious dilemma. Even a short-term solution was better than none at all, so they agreed to try for a depressurization by allowing steam and water to blow down into the torus. The valves required for this activity could no longer be operated from the control room, so the ASEs used telephones to instruct various individuals to activate the valves manually and estimate the proper settings. Four main steam line relief valves were cracked, and as the hot steam gushed downward, pressure inside the unit dropped. But so did the level of the vital cooling water. From its normal 200-inch depth, it fell steadily to a stabilization point only 48 inches above the top of the still-active fuel.

The lower pressures, however, allowed the condensate pumps to force water in, and there was some sign of a depth increase.

But the problem with the torus itself still lingered. Two men were dispatched to the reactor to try and enter the building and work another valve by hand. Armed with breathing apparatus they made three attempts but were stopped each time because they had only 18 minutes of air in their backpacks. This was not long enough to allow them to reach the valve, operate and adjust it, then make their way out of the building. The tanks, designed to contain a much longer supply, were only partially filled, due to low master tank pressure. Dejected, the two returned and reported their failure to the shift engineer.

In the meantime, by using the telephone to contact men at other stations, some measure of order was being restored. Operators working from relayed instructions, were performing required adjustments. Things looked better than they had for some time.

Then the fire struck again. The PAX outgoing telephone system, which had operated even after power in the building had failed, now went dead. It was possible to call in, but the crew in the control room could not dial out with vital value-adjustment settings.

Runners were dispatched to get the teams working the valves to phone in to the master control at regular intervals.

More and more staff members, sensing the seriousness of the situation, gravitated to the main area. Working technicians were harassed by newcomers barraging them with questions designed to gain assurance control would be restored.

Pressure in the giant unit was stabilizing. The new pump hookup was able to add water to the confined space. By 2:00 P.M., the reactor steam pressure had moderated to about 200 psi and the water level was its normal 200 inches.

During this adventure, Unit 2 was having its own little problems. The same irregularities in the monitor panel lights had been going on for some time. The alarm system seemed to be upset, operating independently of any observable reality. Two components of the emergency core cooling system had failed almost immediately and a single remaining set of pumps and valves began to falter on an intermittent basis. The complete low-pressure system was still available on standby. To be doubly sure of not having a replay of the number one reactor, the number two controller set up the same linkage of high-pressure rod operating pumps which had proved successful earlier.

During all this time the fire was being fought erratically. It had spread even farther and more and more wiring was being consumed with every passing minute.

Command of the fire fighters in both the main cable spreader room and in the reactor building had changed several times. One assistant unit operator, who had directed the operation for a considerable period, was finally forced back to the main control area suffering from the effects of the noxious smoke.

By 1:00 o'clock, thirty minutes after the first alarm, the fire-suppression crews had been divided into two operating sections. One, in the spreader room, was directed by an assistant shift engineer. The other, in the reactor building, by a second management assistant. They were not professional fire fighters, and neither had had any special training for this effort. The only help came from instructions they were receiving by telephone from the shift engineer's office. A decision was quickly reached to contact the Athens (Alabama) Fire Department, and by about 1:30 P.M., the fire chief and several men were at the scene.

In addition to a supply of properly filled bottles of air, the new arrivals also had very definite views on how to stifle the blaze. Recognizing he and his men were under the direction of the Browns Ferry personnel, the fire chief made all his knowledge and facilities available to them.

Several relays were formed to run empty air bottles back to the station in Athens for recharging. Additional backpacks were filled by using the pump in the truck. The chief's advice—to quench the fire by spraying it with water—was considered but rejected. The plant supervisor, fearing some of the wiring might still be carrying high voltages, disallowed the use of water and instructed the workers to continue using chemical extinguishers.

About 3:00 P.M., an off-duty shift engineer arrived and immediately took charge of the fire-fighting teams in the cable spreading room. He continued the chemical attack but was able to better direct their application. An hour and a half later, four hours after its start, that fire was reported completely out in his area.

The activity in the main reactor room, however, was not proceeding as successfully. Flames were in the trays high up along one wall, where it seemed nothing was effective in suppressing the blaze. They would die, give off thick, foul-smelling, oily smoke, then renew themselves, burning brightly.

The Athens fire chief argued for the use of water. It was his opinion the fire was not electrical in nature, but was, instead, semichemical, caused by the burning of the insulation. He theorized if they sprayed the cables with water, it would cool the wiring enough so the wires themselves would not rekindle the

160

blaze. There was a great deal of argument, but it was again decided to continue suppression efforts by using chemical extinguishers. Twice rebuffed, the chief was highly irritated but still cooperative.

The main control room situation, although better than earlier, remained bad.

Jury-rigged lighting was being used, as all power had long ago failed. Brilliant puddles of light were crossed through with streaming tendrils of black and white smoke. A team manned the telephone, relaying instructions as each station called in. Residual smoke still clung to the ceiling and the stench of burning cable insulation permeated the air. The scene was more like a battlefield than an orderly control center, but the operators were managing to cope with each development.

Still more unneeded workers had moved into the long room to be at the heart of the action, and they added measurably to the overall confusion. Individuals were coming and going, hurried conferences were held in tight huddles of two or three, and instructions were shouted across the room.

Torus cooling was still a difficulty. The diversion of the condensate pumps to maintain water in the reactor chamber added to this problem. The torus, technically called the reactor containment suppression chamber, is one of the final defenses against radioactive leakage from the core. It is designed as a major component in the system and is a concrete and steel bottle which contains or surrounds the reactor pressure vessel. Should the pressure vessel fail, the torus is the next and final barrier to hold in the radioactivity.

The Unit 1 operator could not tell the water level or temperature inside the unit because all gauges were inoperable. But even with the relief valves working, the residual heat removal system (RHR), which maintained cooling to the torus, was out, so the technicians felt things were not well. Additionally, without the RHR, final shutdown cooling was impossible.

But time had been bought through trial and error. With a shrewd speculation about the extent of the damage, both the water level and pressure in the reactor chamber itself were being maintained. Once assured of this, the operating technicians began sketching possible methods to apply cooling to the overheated torus.

At about 2:00 P.M., the radiation monitors on the Unit 1

161

reactor building had failed, so there was no longer any way of telling whether radiation was leaking into the atmosphere.

The environment control personnel began taking samples on a "grab" or random basis, as an improvised alternative to the continual sampling usually provided by the installed instrumentation. Everyone was relieved when the results proved to be about normal.

At 3:20 P.M., the problem of cooling the torus had become really serious. The State of Alabama Emergency Plan for Browns Ferry Nuclear Plant was implemented to the extent of notifying several designated individuals. This action could hardly be called successful.

The Director for Radiological Health for the State of Alabama was not told of the fire until 3:20 P.M., some two hours after its start. He was called by the Tennessee Valley Authority Environs Emergency Center Director from his office in Muscle Shoals, and given very sketchy information. Aside from a statement there had been a fire and both reactors were scrammed, no details were offered.

At 3:40 P.M., an unsuccessful effort was made to contact the state health offices. One call went unanswered, but no further attempts were made.

At 3:45 P.M., the Director for Radiological Health managed to get through to the Alabama Civil Defense Department and advised them that according to available data, radiation was not above permissible levels. A request for Civil Defense to initiate all emergency plan notification procedures was also made. The on-duty Civil Defense staff member made a desultory effort to do so, reaching only a limited number of the specified individuals. After trying for only an hour, he discontinued the task.

A central emergency control center was established by the TVA, and grab-sampling of the atmosphere for indications of radioactivity in excess of those normally recorded was instituted by TVA personnel.

While all this was taking place, several additional events had occurred in the Unit 1 control center.

Although controls were still out, special switchovers had been made. These, along with some additional repairs, returned a few instrument panels to operation. No control was available over

162

the residual heat removal system, however, so things in the torus seemed to be getting worse.

Constant pumping had completely restored the water level over the fuel core. Various units were able to hold the standard 200-inch level without strain. Realignment of one of the systems for torus cooling was, however, impossible. Several calls were made for the assembly of all available off-duty electricians and operators, and a preliminary plan was completed which called for a major on-the-spot rewiring, even though the fire was still burning and additional circuits might fail at any moment.

By 2:00 P.M. that Saturday afternoon, many people were at work in the big control room, adding to the tumult but re-establishing necessary links. They would work for hours, patching where they were able, before improvements of a substantial nature could be discerned.

The furious fire was still burning, causing more and more havoc, although by about 4:00 P.M., though additional yards of cables and wires would be lost in the blaze, the real damage—to the major control devices—had already been done.

About five minutes after five, the Director of the Environs Emergency Center on the Browns Ferry site gave an order for more air samples to be obtained. Smoke pouring from the reactor building looked alarming and the threat of some atmospheric pollution from the earlier high pressures in the core container caused everyone to move with extra caution. The samples showed no trace of any unusual radioactivity, but as a general precaution the meteorological tower was ordered to be evacuated.

Around 3:00 P.M., the operator in control number one had instituted a simple procedure to bring one of the residual heat-removal systems, not used in the water-pumping operation, up to "on line" status. Voltage was still being progressively lost to more and more of the control boards, and the telephone system remained inoperable for outgoing calls, but the work from the newly arrived electricians and other personnel gave the maneuver some hope of success. To carry it off, one or more men, wearing air tanks for breathing, had to enter the fire zone, pass through it, and manually align the necessary valves. Then, since there was no direct communication with the master control, they might have to go back for minor adjustments.

163

The difficulty was getting there in the first place. The backpacks contained just enough air to last about eighteen minutes—not long enough to move through the smoky darkness, locate the valves, adjust them, and return. Although a similar maneuver had been successfully achieved earlier, the distance from the door to the valves this time was too great. Two teams of two men each made separate tries, but neither was successful.

They returned to the control room to face growing concern. There was no question the torus was operating at or near its basic design limit. Something had to be done.

The fire in the reactor building had to be quelled to allow passage of the crew to the valves. There was no other solution. Several intermediate actions could be taken, but they were mere stopgaps. Without the valves being set, there was no way to avert a serious disaster.

Additional work on the control boards proved effective in aligning the RHR system, but since there was no way to ensure that the pipes were full of water, the decision to start the operation of the pumps was delayed.

Then, around 4:30 P.M., four hours after the candle first set the fire, power was restored to about half the monitoring equipment.

A request to start the reactor building exhaust fan, to help clear the air of the smoke and stench, was approved at about 4:40 P.M., and the operators tripped the switch. The effect was immediately noticeable in two areas. The air in the control room became almost breathable—but the air draft created by the system spurred the fire into new action. After 20 minutes, the shift engineer judged the improvement of the airflow to be less important than the difficulty it was causing, so the fan was shut down again.

By 6:00 P.M., the situation concerning the torus cooling had become crucial. Immediate action was now imperative. Control of the last four relief valves was completely gone, despite the activity of the repair crew, and there was a limit to how much longer the entire system could be maintained by the use of the rod drive and condensate pumps. The fire had to be stopped so workers could reach the controls or heat and pressure would destroy the torus container.

Through all the earlier hours, the Athens fire chief had

164

spoken to one person after another about using water on the blaze. He could understand their caution due to concern about electric shock, but it was clear to him the chemical extinguishing methods which had worked in the cable spreading room would not be effective on the conflagration in the reactor building.

Again and again, his suggestions were rebuffed. The desperate need to pass a man through the fire area, however, finally brought a positive answer. The risk of damage and injury from water being sprayed on live electrical lines was far less than the major disaster which would occur if the torus ruptured.

An operator, equipped with breathing equipment, was stationed at the door of the reactor building. He would need the approval of a shift engineer before entering the blaze area, so a telephone line was kept open.

The fire fighters unreeled a hose and turned on the water. A thin, disappointing stream dribbled out and in the confusion it was decided to replace the nozzle, which was thought to be defective. One was obtained from the Athens fire company. In reality, the hose was only partially unwound from its reel. The slow flow was caused by the tightly wound portion still on the storage rack.

The borrowed nozzle had threads which did not match the end of the fire hose, and although an attempt was made to fasten it in place, the results were less than satisfactory. It came off as soon as the water pressure became strong enough to direct a cooling spray onto the trays of cable running high along the wall.

The problem was finally resolved, and the shift engineer, along with two other men, could enter the area of the main blaze. Climbing to within four feet of the flames with the assistance of two other workers, the shift engineer sprayed the cable tray for only ten seconds. The fire was out. After clambering back down to ground level, he left the hose stuck in a position where it would continue to wash the troughs, then the three retreated.

Meanwhile, another team had moved onto the second level. There, too, the use of water on the fire proved successful. Once the Athens fire chief's suggestion was followed the blaze was completely extinguished in less than ten minutes. An hour later, after long-term cooling by the water sprays, the fire was officially declared out.

The operator stationed at the door of the reactor room moved as soon as it became apparent there was no further threat from the flames. Running, he made his way through the debris to the critical valves and operated them by hand. Joined by others, he soon had the adjustments right, allowing the residual heat-removal system to be operational.

There was still more drama in the control room because the gauges monitoring the water level in the system were faulty, so there was no way to tell if it was full or empty. To start the pumps with air, the pipes could cause cavitation and stop the water flow for hours.

By 8:00 P.M. Saturday night, the electricians had managed to bring the largest number of boards back into service. Another crew of technicians restored the telephone system so outgoing calls could be made. A third team was well into the job of flushing the RHR system.

The problem of pressure in the reactor chamber still lingered. Residual heat in the uranium core was causing the water above it to boil, and the constant addition of steam increased the pressure upward to more than 300 psi. Periodically, to relieve the strain, the operator would order certain valves opened so the vapor would force itself out into the torus. This "blow-down" maneuver was thought up as an emergency measure, and it was effective because it bypassed many of the inoperable systems which would normally have accomplished the necessary relief process in a more orderly manner.

Each time the reactor was blown down, the temperature in the torus increased. Finally, at about 8:30 P.M., the area called the dry well began to be affected by the unplanned-for heat. Pressure built up in this relatively cool zone until there was fear it might rupture or burst. A release was achieved by the unorthodox measure of venting through standby gas treatment lines. Steam fitters manually operated the valves routing steam into the plant stack, where it passed into the atmosphere. Monitoring at the time showed very slight increases in radioactivity, but the level was far below any criteria of danger.

The technical staff, working rapidly, was hooking various switches into newly laid lines. By using a trial-and-error setting technique, control was being regained.

But a problem still lingered. There was no way, without the

166

RHR and the final shutdown cooling systems in operating order, to completely close down the pile. They were holding a monster on a leash but could not make it lie down. For the moment, there was no immediate danger; things were in a state of abeyance. But there was no indication of a final solution either. They couldn't turn it loose, but if they hung on too long, something might happen. That something wasn't very pleasant to contemplate.

The control room was in a state of chaos. Non-essential staff members were moving about aimlessly, getting in the way of the workers and the operators. According to later testimony, supervisory personnel stood next to men who were vitally employed, continually asking for reassurances which no one in good faith could give.

One indication of the confusion comes from the actions concerning the reporting of a red light which served as a hazard warning to aircraft. Located on the top of the plant stack, it went out as a result of the fire. As darkness fell, its absence was noticed by a member of the environmental staff. He decided to report it rather than risk an accident.

Instead of calling the FAA directly, he found a public telephone and dialed the guardhouse. When no one answered, he was even more determined to call attention to the problem so he phoned the Environs Emergency Center, which, finally called the FAA. It took more than a half hour to report a missing light.

Working through the night, the weary men in the reactor control center made progress. With one team monitoring the state of the core and running the necessary blow-down pressure reductions when required, the rest aimed at restoring power and cleaning the RHR system so it could be used in a final reactor shutdown. The control rod pump and the condensate pumps still labored to provide water in the core chamber, and the level was holding well.

Finally, around 2:00 A.M. on Sunday, the torus instrumentation hookup was completed allowing a measurement of the water level inside. It was within its limits and one more worry was quelled.

Work continued through the long night. The RHR system was flushed but could not be put back into use until several additional connections were made. One final period of strain occurred during the closing phases, but enough normal controls

167

had been installed to prevent another full-scale call to action. Finally, at 10:40 P.M., an entire day later, shutdown cooling was attained by the normal flow path and the reactor was in its closed state. Reactor pressures dropped to less than 10 psi, and the monster on the leash had been laid to rest.

Air samples taken by the Environs Radiological Emergency crews, who remained in action until 5:00 A.M. on the day after the fire started, showed no significant radioactive release. The dosimeters and filmbadges worn by the workmen to indicate their exposure had similar negative readings.

A total of seven employees reported to the hospital with varying levels of discomfort from smoke inhalation. Follow-up medical reports indicated none of the men lost time from work, and each came in for his next regular shift.

It had been a close call. How close is guesswork. With system after system failing, the operators and engineers were nevertheless able to maintain a level of safety. The game of "what if" has been played by numerous individuals who write for the antinuclear press, but none of them, apparently, took the time to review the actual reports of the proceedings.

The Browns Ferry fire shows us many things about our nuclear power generating facilities. And not all of them are good.

The people who work in the plants performing maintenance and repairs are often not sufficiently informed about the problems they can cause. And management is too lax. On the Thursday before the disastrous blaze—and it certainly was disastrous financially because in the end it cost over ten million dollars—there was not one, but two minor fires caused by the same flickering-candle leak-detection technique. They occurred in cable penetrations similar to the one which got out of hand. The first little incendiary incident was put out by an electrician with his fingers. The second lasted only about 30 seconds before being snuffed out by a CO_2 extinguisher. One incident was reported to construction supervisory workers. The other was entered in the shift engineer's log.

So there was certain knowledge of the possibility of a fire from the technique in use. But no one did anything about it.

A second problem lies in the design of Browns Ferry and some other plants. There are supposed to be two totally separate, completely redundant control systems. But this proven space-age technique is defeated when all the cables for all the controls pass through a single area. Why, then, is this allowed to happen? From a construction standpoint it is easier and far less expensive to design a plant with a just one cable-carrying area. But as is now easy to see, the simplest solution may not always be the best one—if a design flaw like this defeats the redundant backup system, no amount of financial savings can possibly justify allowing such potential for disaster.

The cable trays themselves, which provide no effective fire barrier, had been the subject of several NRC memos. These same documents indicate the NRC was concerned with the commonality of the wiring for both primary and backup systems. But no action was taken.

Even though the operations and engineering staffs performed brilliantly under real stress, they were not perfect. There was a completely different hookup, joining the number one and number two control rod pumps, which escaped everyone's notice. The men on duty would have breathed easier if they had known of this backup. And if the single high-pressure pump had failed, their lack of this knowledge would have resulted in a disaster of uncertain, but undoubtedly great magnitude.

Another difficulty lies in the general lack of preparedness on the part of those assigned to the various functions of the emergency plan. From the fire fighters to the officials at a state level charged with conducting certain citizen-protection operations. There was a visible deficiency in training and planning.

In the volumes of documents produced by the NRC during its investigations of the fire, and in the records of the Congressional hearings which touched on the matter, one point is repeated many times in the testimony of individuals responsible for the functioning of the emergency programs: the plan was poorly implemented.

The fire fighters were not trained for fire-suppression work, and the superiors responsible for directing them did not even know, at times, the extent of the blaze.

TVA's Central Emergency Control Center (CECC), responsible for executing the emergency plan, was cited in the NRC report as "not being well coordinated." CECC communications with the Browns Ferry site were characterized as "not effective in keeping the CECC currently informed" and "communications with other

169

agencies [the Alabama State Civil Defense and groups responsible for emergency action] led to misunderstanding of plant status by those agencies."

The individual who would normally have functioned as director of the CECC was unavailable when needed. His replacement did not know this and worked at the CECC for half an hour before finding out he was the man in charge. As late as 5:00 P.M., the CECC was informing other agencies the fire was confined to the cable spreading room. And the CECC director did not know the fire was out until an hour after the fact.

The TVA people were not alone in their ineptitude. According to the NRC report, Alabama's emergency plan was not available to certain participants, individual responsibilities were not clearly assigned, and the plan was actually out of date. Even worse, attempts to contact city and county officials by state authorities was minimal.

From the very start, safety procedures and precautions were not followed by the TVA. A one-page document dated February 11, 1972, and used by Browns Ferry personnel to learn how to report a fire, contained a gross error. The emergency procedure specified a person discovering a fire would:

"1. Sound fire alarm and report condition to control room operator. (Dial 299 and wait for an operator to take information.)
"2. Take immediate corrective action (use fire extinguishers, etc.). If not sure of correct action, wait for instructions.

"Dialing 299 automatically activates a plant-wide fire alarm and rings the telephone in the Unit 1 control room."

The listing of numbers in bold print at the bottom of this same training document, however, did not show 299 as the correct fire number. Instead, it indicated 235. This poster was displayed in a number of areas inside the facility as an emergency quick-reference guide, confusing the guard who first tried to report the fire.

Worse, a plant Standard Practice Manual agreed with the 235 number but made no mention of the plant fire-alarm number 299, or the need to inform operating personnel.

Further references to the NRC report show the sheriff of Morgan County was notified, but "had not been briefed in the State of Alabama Emergency Plan," and he did not have a copy of the document. But he fared better than the sheriff of Limestone County, who was never officially notified at all.

The problem of insufficient attention to the emergency aspects of the situation, combined with a general lack of knowledge of necessary procedures, is documented by pages of testimony and reports. One defense has been offered, centering on the idea the situation was not sufficiently serious to merit full implementation of the disaster procedures. According to this viewpoint, had the problem been truly serious, better action would have been taken.

This stance is hard to reconcile with the facts. The individuals in the control room considered their problems to be very real—and of a magnitude capable of producing a disaster if they were unable to cope with the loss of equipment. The supervisor who asked continually if things were going to be all right was a worried man. And there were a lot of other worried men in the room on that long afternoon.

There is a vast difference between the cool historical reporting to be found in the records of the Browns Ferry fire and the atmosphere of emotional upset, trauma, suspense, tension, and confusion in the main control area during the incident. Laid out with precision, the events tell a logical story. Mistakes are merely recorded among the other facts. Even the style of writing the reports is, as it should be, unemotional and formal. But it is not hard to read through the maze of material and see the underlying human drama taking place. There was no shortage of fear at Browns Ferry on the day of the accident.

On the afternoon of March 22, 1975, we came close to a real disaster. How close, no one can really say. Those involved performed as people always do in periods of dire stress: some do well, others poorly. And postmortems rarely achieve more than a factual outline of what happened. They offer information for technical improvement, but they do little to prepare for the next round. The next time, other things break down. The next time perhaps Murphy's Law comes into play.

■ Not all nuclear plant designers still cling to the "totally safe" syndrome which was the rule in the halcyon days of the 1950s and 1960s. There is a new way of thinking nowadays. Additional knowledge, has, in many cases, brought forth an understanding of what is not known—and, better still, a questioning of what had been heretofore accepted as safe standards. If the examples reviewed so far indicate nothing else, they point inescapably to the fact that nuclear accidents can and very likely will occur. It is not a question of *if*, but of *when*. And *where*. And *how bad*?

To date, the industry has a remarkable safety record. Not a single individual outside the atomic field has died from radiation exposure.

But the scientific community is not sleeping on the subject. Dissident voices are increasingly being heard, and these voices are often in exact opposition to the often almost platitudinal litanies of the upper-level management of our nuclear programs.

The NRC itself houses one such new wave of thinking. Top scientists and engineers are bypassing normal channels to come to the front in order to speak on safety and the laxity of standards and safeguards.

In the field of nuclear power generation there are presently 65 plants operating in the U.S. alone. More are scheduled to be built in the next few years.

But secret studies, which have surfaced through the activities of a number of the concerned scientists, indicate the return on investment of electrical power companies in this field may be too low to allow for the long-term repairs and maintenance to keep them at maximum safe operating levels.

Declining maintenance is not one of the probable causes of a mishap included in the now famous *WASH-1400* study which indicates the probabilities of death or injury related to the atom-powered generators. But a lessening of maintenance standards can and will lead to failures.

The utilities find themselves in a difficult situation. The already committed investments in electrical transportation are huge. Wires, grids, transformers, relay stations, and cross-country systems have cost billions of dollars. They exist and are the logical network to carry power to our homes and factories.

The energy fed into this network has to be generated at a cost which will allow it to be sold without enormous price increases. The atom is the reasonable answer. And the utilities have been a prime force in pressuring the government into moving in the nuclear direction.

This would be fine—except the need for rapid action has caused certain strains in the development system necessary to produce the hardware. The effects of these strains are, for the first time, surfacing.

Published reports indicate two-fifths of the 25 NRC experts charged with the engineering of electronic controls used in power station operation now feel there may be hazards in the systems they have helped design.

This "palace revolt" within the branch of the NRC devoted to electrical instrumentation has reached the point at which individuals have resigned to be able to present their views in public.

The situation was serious enough in late 1976 to encourage Bernard Rusche, the NRC's Director of Reactor Regulation, to send the Commission's Inspector General, Thomas McTiernan, to consult with the dissident specialists.

The result is said to be a report of about 300 pages. Unofficially classified since its submission, its full contents have never been made public, but a special summary has received some limited circulation.

The dissident members seem to feel the NRC has been too close to industry for its policies and policing to be effective, and there is an all too real probability of critical apparatus failure.

One specific point which has come out of this group's discussion of the subject before a meeting of their technical peers is their disapproval of the design of a redundant auxiliary control room. In itself, a second command area is a meritorious idea and would maintain operator control of an installation even if the main room had to be evacuated in an emergency situation. The problem, however, stems from the design of this standby system. According to plans presented, it would be coupled to the main instrumentation and could operate the same equipment by remote control. The standby room, then, is not a fail-safe measure and would itself not function if there was damage to the equipment in the central control area. It is likely many contingencies serious enough to cause the evacuation of the main room would result in some damage to at least a portion of the operating equipment. With these master controls gone, there would be inoperative systems and the standby site would be unable to function. This could well lead to a disaster. Electronic wiring responsible for the operation of the plant is also redundant, but the Browns Ferry incident showed the folly of using common pathways through the building.

The same dissident group also points out the amount of time between loss of coolant in the system and the attainment of an irreversible point in the heating of the pile, which would lead to a meltdown and the subsequent discharge of radioactive pollutants into the atmosphere, is an unknown.

Theoretical estimates of the time available before the point of no return run from a few minutes to mere seconds. Research is progressing to gain a more accurate calculation, but clearly the operators need to be backed by redundant fail-safe systems. The NRC maintains they are. Some of their staffers and other highly qualified observers question this.

The issue is more confused by our increasing knowledge in the field. As our systems technology improves, it is offset by our growing understanding of the nuclear process. In other words, as we learn to deal with one contingency, another is discovered.

Bernard Rusche, in a lengthy explanation of the NRC's present position, stated there has never been an instance in which private industry was favored at the expense of plant safety. "If the plant's not safe," he is quoted as saying, "it's not going to operate." But the definition of the word "safe" changes as both our knowledge and technical ability grow. The dissidents claim a plant which Rusche says was finally shut down permanently had operated the twelve previous years without proper safeguards.

Certainly, in the case of the Cimarron facility of Kerr-McGee, the operation, by the very nature of the process techniques, exposed many workers to contamination.

Today, workers face less likelihood of exposure to radiation. But the present systems are still lacking.

The question which arises from all this is a difficult one to answer. In a country based on private enterprise, the business sector of the community is expected to develop the power sources, while the government is cast in the role of public watchdog and is in the position of establishing controls, regulations, and checking procedures to assure compliance.

The government cannot ignore the needs of businesses to realize a reasonable return on their investment. Yet in its larger role, the government is responsible for the maintenance of the society's standard of living. It must be alert to changes which could adversely affect the economy. Energy supply is one of the greatest of these possible alterations. This results in a circumstance in which the government is forced to set standards allowing the private-investment sector to develop and build atomic processing and power facilities to meet the nation's growing energy needs, while at the same time exerting controls on these facilities which will be attainable by industry and business at cost levels giving some reasonable return on their investment.

It is a hard task, and those who are charged with its accomplishment walk a tight and narrow line between safety and practical need.

Government-owned and operated nuclear facilities are subject to their own problems. While federally developed atomic power plants are not so subject to the pressure of producing profits, other difficulties enter the picture. For some reason, one governmental office has problems in enforcing its regulations on another. The results can be chaotic.

12

■ One of the most unique things about the United States, since almost its earliest days, has been the nation's preoccupation with roads. From colonial times, all-weather highways have linked our most important population centers. Our country is large, and as we grew westward the need for more and more miles of roadway became pronounced. Names like Cumberland Gap, El Camino Real, Wilderness Road, and the Overland Trail are a part of our history.

The real impetus for the development of a superior network of well-surfaced roads came with the advent of the automobile. In a fifty-year period, from 1910 to 1960, our country constructed more miles of new highway than any other nation in the entire history of man. Some think this may not have been for the better. But sociological arguments aside, the development of the freeway and the perfecting of the internal-combustion engine gave rise to a new transportation industry.

During World War II, the "Red Ball Express," charged with supplying General Patton's armored divisions with fuel, food, and ammunition, proved the efficiency of rapid motor-vehicle transport on a scale never before attempted. By the close of 1945, companies in the U.S. were eager for the opportunity to duplicate portions of the "Red Ball" operation on our own blacktop thruways.

Special truck bodies were developed to haul everything from furniture to oranges—or gasoline and toxic chemicals.

The nuclear industry has a real need for surface transportation, especially for its raw materials, fuel components, and waste. It is far better to move a dangerous substance cross-country at ground level than to add to the risk by flying it.

There are problems even with surface transportation, however. Standards are somewhat confused, and criteria for safe containerization of radioactive waste are a little vague. Worse, many of the shipping companies called upon to move these nuclear cargoes are not aware of the possible consequences of even a minor accident.

This next story is not atypical. Incidents like it have occurred many times. To date, no one appears to have been injured. But it is one more example of our overly casual attitude toward the by-products of our nuclear time. And any incident of this type could have a truly tragic ending.

■ Highway 287 leaves the town of Limon in eastern Colorado paralleling the banks of the twisting Big Sandy River through the settlements of Wild Horse and Kit Carson. Then it turns straight south and with a few jig-jogs runs through several Indian reservations, crosses the mighty Arkansas River, then dumps into Oklahoma.

The country bisected by this glistening blacktop road is some of the most fertile in the United States. Rolling in parts, it lies north of the highest point in Texas but far east of the towering snow-peaked Rockies.

Trucks travel this stretch in great numbers. The distances between the small cities and the lack of rail freight facilities make motor transport the standard means of commercial communication. Most of this road is easy driving unless the weather is bad. Then, blinding blizzards of blowing snow can stop all traffic for days, despite the best efforts of the state and county crews.

September 27, 1977, arrived during one of the good-weather times. There was no sign of rain, sleet, snow, fog, or dust. And the wind was even down, which is a rarity thereabouts.

Traffic was light, due more to the hour than anything else. A lone truck thundered its way along the two-lane blacktop. The driver, Donald Atwood, stretched in his seat to relax stiffening muscles. The cab-over-chassis he was pushing to tow the rest of the 18-wheel rig was a late model. Even at top speed it rode well. Wind noise, although present, was sealed outside, so conversation inside the cab was easy. Don turned slightly in his seat and in the reflected flow from the dash instruments he could see the profile of his companion and relief driver, Willima Pipher.

Trucking had changed in the last few years. The presence of women who rode as both lead and backup jockeys had somehow softened what had by tradition been an all-male industry.

178

Don scanned the night. His huge headlamps punched holes in the dense darkness of the early morning, sweeping the roadway ahead with an intense brilliance. The clock on the dash read 1:00 A.M. on the nose. They'd stop soon, for a stretch and a cup of coffee. Might be a good idea to let Willima take over for a while and sack out until just before dawn. That was the time driving was the most dangerous. More people starting on the road, and visibility in the not-quite-night, not-quite-day was always at its worst. Don's thoughts wandered slightly, then came back to match his attention, which had never varied from scanning the road unwinding endlessly ahead.

The early hour of the morning, that desolate time before dawn, had been selected as a safety measure by the manufacturers of the cargo stored in the big van. Fifty-five-gallon steel drums were stacked from floor to ceiling, front to back, filling all the available space. Inside each of the round metal containers was a fine, yellowish powder. In all, the truck contained 40,000 pounds of the material known as yellowcake. Produced from uranium, the radioactive substance was fodder for the nuclear industry. Further processing would develop it into a more refined fuel, much of which would finally end up inside a reactor in an atomic power station.

Shifting again in his seat, Don Atwood watched the night roll by. Nothing seemed out of the ordinary. Then, without warning, what appeared to be an entire herd of horses, manes and tails flying, erupted into the white flame of the headlights. Startled, they ran diagonally across the road, directly into the path of the oncoming truck.

Atwood swung the big steering wheel sharply, cutting to avoid the inevitable collision. Later, he explained his stunned shock at the appearance of the ghostly animals: "All of a sudden there was a whole herd of horses crossing. One minute they weren't there and the next minute they were."

The truck, wrenched from its straight path, resisted the turn with several hundred tons of inertial energy. Then, tires screaming, it veered across the road. Reacting coolly, Atwood compensated for the skid, steering quickly into the new direction of travel. Then it was over. The huge vehicle had run out of roadway. Coming back to the centerline, the front edge of the multi-ton truck struck three madly running horses, gave a lurch from the force of the impact, and swerved sharply to the right. The edge of

179

the narrow highway was graveled and the now out-of-control 18-wheeler flashed over this verge and slammed into a narrow ditch, plowing dirt and sending a spray of gravel high into the pre-dawn air. The big truck bumped and slammed its way forward. All control lost, it overturned with metal tearing from the chassis in irregular slashes. The top of the cab was battered down on the two occupants.

In the van, the heavy barrels of yellowcake erupted. Bursting through the aluminum walls of the vehicle, some flew through the air for yards before smashing into the earth and splitting open, spilling their contents onto the barren black ground.

Finally, after a seemingly interminable period of screaming noise, what was left of the big truck came to a final stop. Ripped apart by the impact, the remains of the vehicle lay in the middle of a small field of spilled yellow powder.

The silence after the rain of sound seemed almost as loud as the noises of the smash itself. Dazed, Atwood, who was squashed down into one corner of the cabin, looked around.

Everything had been forced into crazy angles so it took him a minute to get his bearings. Willima was there, moaning softly, injured by having been thrown around inside the cabin.

Insect noises resumed and the night sounds filtered lightly through the broken glass of the battered vehicle. Softly on the wind the smell of fall and a promise of coming cold wafted through.

Atwood tried to move. He was pinned into a small area. The crumpled steel around his head formed a prison. A moment of panic ran through his mind as he thought of the fuel spilling out and the possibility of a fire. He calmed himself by trying to devise some way of freeing Willima.

The highway remained deserted. Quiet, broken only by an occasional tinkle as a shard of smashed glass freed itself and fell onto the rest of the wreckage, was a presence over the land.

Up the road, moving slowly, another trucker sat humming to himself. The lateness of the hour lulled his mind, forcing him to continually pull his attention back to the slashing corridors of light. Then, suddenly coming totally alert, the man started. Eyes staring, he watched the scene of the wreck grow closer. The broken truck, one headlight pointing across the field adjoining the

180

highway, was outlined by the approaching vehicle's lights. There was a squeal of brakes as the semi came to a stop, then a revving of the engine as it backed and turned to allow its many lights to illuminate the scene. The driver sat stunned for a moment. Then, switching off his engine, he climbed down and started for the wreckage. Walking was difficult, as there seemed to be some kind of yellow powder almost ankle-deep everywhere around the demolished trailer.

Moving in close to what appeared to be the front window, he used his flashlight. The jumble inside took him a minute to sort out. Then he caught sight of Atwood's staring eyes. Realizing there was nothing he could do alone, the man muttered a few words of encouragement and then ran for the radio inside his cab.

Don Atwood and Willima Pipher had been inside the broken wreckage for almost fifteen minutes.

Alerted by the driver's single sideband announcement, truckers farther back on the highway relayed news of the disaster to a local base station, where a call was placed to the highway patrol. By the time they arrived, forty minutes after the crash, several additional trucks, other automobile drivers, and a wrecker were on the scene.

In all, according to police estimates, over sixty people stopped to see if they could render assistance—and each of them marched through the yellow radioactive dust as work progressed on the project to free Atwood and Pipher.

The investigating officers followed the usual accident procedure. Required medical and mechanical assistance was sent for. A report was made up, but could not be completed until the driver could be questioned. No one, moving in the floodlighted darkness from areas of dazzling brilliance to black shadow, took much notice of the spilled material.

Yellowcake is not a highly toxic substance but it is dangerous if inhaled. And even though some containers had remained intact, about 10,000 pounds had spilled onto the cold ground. Another 5,000 pounds was loose inside the twisted van. All told, there was an ample amount of the fine powder available for inhalation.

Working as rapidly as the conditions of darkness would allow, the various assistance groups managed to right the truck

181

and free the two captives. This operation took the better part of three hours, during which time more people came and went along the still open road. Finally, with Willima Pipher safe in medical care, the officers were able to ascertain the nature of the spillage.

With no one particularly alarmed, a routine report was filed. None of the people on the accident site wore respirators to filter the dust out of the air they were breathing. But the officers were sufficiently concerned to get the names of the 26 persons who had assisted in some fashion during the long hours required to release the two drivers.

In the meantime, cars continued to pass. As the night slipped into early dawn it brought lower temperatures. Most of the pass-ersby had their heaters on. The light wind which stirred the dust could easily have blown some into the air intakes of the moving vehicles.

In the first light, the eerie-looking yellow powder was seen to be completely covering an area of about 30 by 50 yards. Foot-prints through the material, as well as places where it had been trampled into the soft earth and was now streaked with black, showed the passage of many people. Cars still passed as the dim dawn became bright morning. Split, broken drums littered the area, and the battered body of the truck cab, wheels hanging at odd angles and the name LEE, still readable, tilted toward the ground. The scene resembled the aftermath of a bombing. Uni-formed officers were moving about, completing their reports and showing new arrivals the problem. Twelve hours later state health officials would order everyone in the area to wear air filters and the police would set up roadblocks to detour passing motorists from the spillage.

During this time, a number of people were contacted: the trucking lines; Exxon, which owned the cargo and was having it shipped to Oklahoma for further processing; the state health per-sonnel; the state police; and others. A conflict arose over attempts to assign responsibility for the accident and decide who should be charged with the job of cleaning up the scattered mess. Some finger-pointing went on as each party stated its limits of liabil-ity. A full day was to elapse before Exxon agreed to undertake the decontamination process and restore the yellowcake to its containers.

On September 28, the day after the accident, an Exxon

industrial hygienist arrived on the scene. According to reports, even though the wind had increased its intensity slightly, no air samples were taken to determine if some of the yellowcake was being blown away. After a survey of the site, the hygienist called for various forms of assistance. Exxon sent Geiger counters, expecting to use them to determine the levels of radioactivity present.

Shortly before this, the Colorado state officials had already run tests of their own. The normal background level of radiation for the area was about 15 to 20 microroentgens per hour. In the heart of the spill, the count had increased by 4,000 percent to more than 800 microroentgens. While this level is not directly injurious, it is a quantum leap upward, and there is great controversy about the prolonged effect such levels of exposure might have on an individual, even if he did not breathe the material.

After the reported arrival of the Geiger counters for use in detecting the radiation levels, the state health authorities had to explain these instruments were virtually useless for measuring yellowcake radioactivity, and the more sophisticated gamma scintillometers and alpha survey devices would be required if any readings were to be taken as meaningful.

Further hassles occurred when Exxon stated its desire to hire ordinary day labor to perform the cleanup task, instead of trained health environment workers. In the initial stages, shovels were used to scoop up the powder, even though the wind had increased in force, and according to reports, none of the crew were issued respirators.

Again the Colorado health authorities intervened and Exxon issued the protective equipment.

Working long hours, the men slowly managed to scrape up the yellow powder from the ground. Using black industrial plastic to wrap and cover newly delivered 55-gallon drums, they filled them one by one.

But not all of the material would ever be recovered. Blown away by the wind, and perhaps sucked up by passing cars, a measurable amount was lost into the environment forever.

Seven days passed before the public was made aware of the mishap. Even then it was revealed only because a hard-working reporter for a Boulder, Colorado, newspaper, Robert Ruby, did the necessary investigation and put the facts together. In his story, the health officials merely urged any people who had stopped at the accident site or had driven by the scene to "wash their clothes and shower to wash out any contamination."

The matter did not end with the cleanup, however. Two members of Congress reportedly joined with several groups to file petitions with the NRC calling for all licensees which transport radioactive materials to have emergency units for use in case of accidents. At the time of this writing, the matter is still pending.

Several groups have claimed this incident and the reaction to it is another example of how the antinuclear forces will develop any situation into a major catastrophe. They point out, and rightly, that yellowcake, although radioactive, is not generally a material which might be considered a health hazard—not nearly so much, as, say liquid ammonia or other caustic or potentially explosive chemicals shipped through heavily trafficked cities on a daily basis.

But certain things are clear. More care is needed in the transport of a number of substances. Nuclear materials are included in this group. The use of ordinary 55-gallon drums which clearly will not withstand the rigors of a wreck should be closely examined. Members of the Department of Transportation and the National Transportation Safety Board reportedly did not arrive on the scene at any time during the accident investigation or the decontamination process. And NRC inspectors were not present until seven days after the incident.

Robert D. Siek, Associate Director of Environmental Programs for the Colorado Office of Health Protection, best summed up the matter in a strong letter to the NRC dated October 3, 1977. If, he wrote, the accident, "had occurred near a watercourse, in a metropolitan area, or in a mountain area, the problems would be compounded many fold."

This affair serves as one more striking example of the low level of preparedness the nuclear industry and most state governments bear down on even a minor-level difficulty. The attitude which seems to prevail is that nuclear materials are just another of the many hazardous substances with which modern industry deals on a daily basis. As stated earlier, there are worse things in general circulation. But atomic materials require different and more careful handling. The harm radiation can cause is not immediately in evidence; therefore, out of mind. Better levels of preparedness are not only necessary, they are mandatory if industry is to fulfill its responsibility to society.

Transportation accidents involving critical materials are not uncommon. There have been almost 250 in the last ten years. Some, like the Rocky Mount, North Carolina, incident in which a truck carrying low-level wastes from a nuclear power station collided with another vehicle, are more serious than others. In this instance, the radioactive substance was in the form of a glasslike resin, inside special shipping containers. But the interesting thing is the truck had come all the way from the Vermont Yankee nuclear plant to a temporary disposal site in Barnwell, South Carolina. This is a long trip and the wreck was the second suffered by a vehicle from the Vermont location. A spokesman from the Yankee facility is quoted as saying "Traffic accidents" are an "inevitable result of the transportation of nuclear material."

This is an obvious statement. It is equally obvious more people in the business should be aware of this simple, yet basic fact—and be a lot more ready to deal with the problem when it occurs.

13

■ One of the most unusual controversies to arise from the nuclear question focuses on the issue of civil rights. Does plant management have the right to subject employees or visitors to even a cursory search? What about a thorough shakedown which might involve probing into body cavities? Does the operator of a facility have the right to enter a home and scan for radiation? Or to have a worker followed, or to require a detailed itinerary of an individual's movements so a number of places might be monitored? To what extent may a plant use armed force to defend itself against a takeover? And finally, has the nuclear industry been involved with law enforcement officers in an effort to maintain surveillance over the more vociferous opponents of further nuclear power?

These are tough questions, especially in the United States, where a strongly developed sense of the right to privacy and freedom from infringement of individual liberties is ingrained in the population.

Arguments about the eventual need for the nuclear establishment to set up a semi-police state in order to maintain security at its facilities have some basis in truth. The existence of plans to block certain highways and re-route traffic during times of potential danger does put into the hands of plant security personnel a certain amount of control over the general public. And the existence of these security forces, armed with automatic weapons and trained in combat techniques, is tantamount to the development of a new and quite formidable private police force.

The relationship between state government and large-area suppliers of electric energy is well documented. Lobby activity for the utilities is commonplace at every level, and political contributions have been made to selected candidates from city councilmen to members of the U.S. Senate. Nothing is inherently wrong with this. But it illustrates the closeness which exists between the nuclear power industry and the structure of our government. Realistically, a certain amount of this involvement is a good thing.

But is it possible to misuse this relationship? Could a vocal adversary to a new atomic installation be noted and singled out for special investigation and attack? Or would it be possible for a company to have a union figure unduly hassled by using the structure of existing laws and regulations?

This is an interesting question—and a hard one to answer definitively, because many of the issues are still undergoing development as the security measures at the nuclear facilities are refined.

But there has already been an instance of harassment of a nuclear adversary by police at the state level. The individual involved could hardly be described as a radical—or even as an avid enemy of atomic power generation. He was involved in no protests or demonstrations. Instead, he was singled out for special attention after he availed himself of his constitutional rights and appeared before a local city council to give his opinion of a proposed new plant site.

It is a strange story and was revealed only through a quirk of fate. Under different circumstances a great deal of damage might have been done to the career of a respectable citizen, and the individual who had been attacked would never have known the role played by the state police in his misfortune.

One example can't embody all the issues or answer all the questions of civil rights and nuclear power. But it does show what can happen when overzealous people are prodded by a series of circumstances to take what seems to them reasonable and responsible actions—which simply happen to infringe on our civil rights granted by the U.S. Constitution.

It will be a long time before the civil liberties aspects of the nuclear age are fully revealed. What appear now to be dangerous conflicts may be paper tigers which fade away with experience. Or the threat which many people keenly feel exists may turn out to be more than just a vague fear. We might be confronted with a harsh new reality.

The story of Robert Pomeroy isn't as impressive, in some ways, as the others in this book. No bombs fall, nothing blows up, no technicians go mad trying to stop a disastrous core melt. No one is even threatened with exposure to a radioactive substance. It is distressingly simple. One man chooses to speak out and is attacked by a small part of a very large state government.

■ January is a cold month even in a place as far south as Dallas, Texas. Situated on the edge of the Great Plains, the city is unprotected by natural terrain from the bitter, cold blasts of icy air referred to locally as "blue northers." As one Texan put it, "The only thing between Dallas and the North Pole is one lone barbed-wire fence, so the wind blows right down with nothin' to stop it."

The morning of January 14, 1974, found the city in the grip of freezing temperatures, and the stifling atmosphere of the city council chambers was a welcome relief from the chill, windy streets.

The meeting had been convened as usual, but there was an air of expectancy in the crowd. Every indication showed this session was going to be far removed from the usual humdrum ritual.

The chambers filled early. People found seats and watched as, one by one, the city fathers and the mayor took their appropriate chairs on the dais. Then another group, clearly together, entered. It was composed of several well-dressed men armed with briefcases. Some bore themselves as if they were in a court of law, speaking with careful formality. Mixed in with the attorneys were a few more relaxed individuals who were along as observers and expected to take no part in the proceedings.

A third bunch, made up of a number of interesting-looking men and women, grouped together on the other side of the room. They talked among themselves and laughed lightly to dissolve their obvious tension.

The final councilman, rising from where he had been seated talking to a constituent, made his way to the podium-type desk, and when he was settled into place, the meeting was called to order.

The early business moved swiftly and the audience followed the various debates with only moderate interest. Finally, with most other matters disposed of, the agenda called for a statement to the council by a concerned group. CASE, Citizens Association for Sound Energy, had few members. About 40 people banded

189

together to oppose the construction, by a combination of electric utilities, of a nuclear power station near the Texas town of Glen Rose. The facility, tentatively named Comanche Peak, was in the final planning stages. The AEC construction permit appeared certain and only a few public and private hearings remained to be held.

The CASE spokesperson, selected as the leader, was Robert Pomeroy, one of the founding members. Of medium height, he dressed conservatively with his hair in a carefully razor-cut style.

Bob Pomeroy was a reasonable man. Not interested in rabble-rousing, he used CASE as a platform to publicly introduce a number of arguments against the construction of the Comanche Peak plant. His appearance before the council had been publicized and a small number of people were interested in hearing his arguments. Pomeroy was known to have a deep concern about atomic energy, as well as a strong opposition to the need for a plant in his area of the state.

A resident of a small suburban community, Farmers Branch, he worked for Continental Airlines as a flight officer aboard 727 aircraft. He had been a pilot for several years, having first completed a tour of duty in the Marines where he had attained the rank of captain flying single-engine jets. After his service he became a third officer aboard Continental's larger planes. His advancement with the company had been on schedule, and now, as copilot, even though he had some years to go before becoming a first officer and earning his fourth stripe, he was happy in his work. Flying was his life. He could imagine no other employment half so satisfactory—or one for which he was better suited by a combination of sanguine temperament and relaxed style during otherwise tense situations.

When recognized by the council, he began to speak in an evenly modulated voice. Dealing with what he felt were the facts, he introduced telegrams from individuals he described as experts in atomic physics: Drs. David Inglis and Henry Kendall, of the University of Massachusetts and the widely renowned Massachusetts Institute of Technology respectively. With the council's permission, Pomeroy read the messages and they were included in the minutes of the meeting. The underlying tone of the comments dealt with the damage which might be caused to the environment by radioactivity and other releases from a nuclear power plant.

Some debate ensued but the meeting proceeded on an orderly basis. As the discussion grew heated, Pomeroy became more intense, but he conducted himself in a professional and direct fashion. At one point, shortly after the crowd had laughed at one of the council members' remarks, Pomeroy offered to debate W. W. Aston, vice-president of Dallas Power and Light, at his convenience. He suggested a meeting place, the First Unitarian Church, but received no acceptance of his challenge.

After his speech, several people came up to Pomeroy and expressed their support of his views. A few dissident voices were also heard, but discussions remained calm. The CASE people, aware of the public interest they had generated, were reasonably happy with the outcome, but knew, as Pomeroy had said, there was really little they could do about the plant with their lack of membership and funds.

The meeting continued after the pro and con groups had completed their presentations, but a large portion of the audience wandered off rather than sit through the droning dullness of further considerations of new street-paving and sewer-repair projects.

As far as Bob Pomeroy was concerned, the affair with city council had come out about as expected. The CASE arguments had been courteously heard, but little had been achieved. Resolved to continue their fight, the members began making plans for the next big showdown. A public hearing by the AEC was scheduled for late July in the Glen Rose High School auditorium. The outcome of this open session would decide whether or not a limited permit would be issued allowing the Texas utilities to start construction and site work for the proposed Squaw Creek reservoir, a vital part of the new plant cooling system. Hearings of this nature had been held in other parts of the country, for other plants under consideration, and local activitist groups had managed to drag the sessions into year-long meetings, thus delaying the actual start of work. Robert T. Martin, a vice-president of TESCO, one of the three utilities involved in the Comanche Peak plant, said he thought the Glen Rose session would take about two days. Pomeroy and his fellow CASE members were trying to develop a clear case to present to the AEC representatives.

Miles away, in Austin, the state capitol, other work was started on Pomeroy.

Due to his appearance at the Dallas city council meeting, Robert Pomeroy's name was included in a list of possibly suspect people. Collected on a routine basis by the Texas Department of Public Safety (DPS), the persons named on this list were subjected to a thorough investigation. A special unit of the DPS had been formed for this purpose.

An intelligence agent, David A. Dimick, was assigned the task of making a report on Pomeroy's background, friends, and present activities. The work took some time and utilized at least two special "informants." Their comments and testimony were included in the document, which was signed by David A. Dimick, agent #2449, Texas Department of Public Safety. Before being appointed to the intelligence division, on July 15, 1971, Dimick had served with the department as a highway patrolman. Neither his work background nor his formal education had prepared him for the role of intelligence investigator.

The report was explosive. It stated that in addition to Pomeroy's having been present at the city meeting to speak out against the plant, he had also been seen talking to a person who had "been a long-time Socialist party organizer in Dallas." Worse, one of the two Ph.Ds who had sent telegrams, Dr. David Inglis, was said to have been mentioned in a 1961 House Un-American Activities Committee study on "possible Communist infiltration into antinuclear energy front groups."

Other comments included specific mention of the First Unitarian Church in Dallas, the place where Pomeroy had suggested the debate be held, as having sponsored a number of radical groups including the Dallas Peace Committee, the United Farm Workers, and Gay Liberation. The report singled the church out for having been host to a "social" workshop in January 1974, where all major subversive groups in North Texas had set up information booths.

The report is said to have read, in part: "The subject came to the attention of this service on 1-14-74 when he spoke at length to the Dallas City Council in an attempt to block the building of a nuclear power station at Glen Rose, Texas, by the Texas Utilities Co. The subject alleged that he had formed a group that would file suits and attempt to stop the building of a nuclear power station."

192

The two Ph.Ds Pomeroy had used as references were labeled "alleged experts," and the overall tone of the document could be taken to imply the "subject" had close connections with Communists, Socialists, far-leftists, or all three.

A special code was used after each item of information as a device to indicate the relative "reliability" of the information. Some was rated as being highly trustworthy, some of dubious or questionable origin.

Once the report was complete and distributed internally through the Texas DPS system, the matter should have been closed. But Dimick, acting on his own, according to later reported comments made by the then-director of the DPS, Colonel Wilson E. Speir, made one last distribution. He sent a copy to the security department of Continental Airlines.

The Continental security staff was dismayed to receive the document. First of all, they felt what Pomeroy did in his spare time was his own business—as were his political views. He had a proven record of stability as a flight officer and there was no indication in any of his files he might be irrational enough to fly a plane into a nuclear reactor. After a brief discussion of the matter, the security staff transmitted the memorandum to Robert Pomeroy's immediate superior in Dallas. No action was specified.

The whole affair would probably have ended there as an aborted attempt by an official state agency to cause trouble between an individual and his employer. But an extra twist was added. The superior to whom the document had been sent was a close personal friend of Pomeroy's. After some deliberation, he decided to show his friend the report and allow him to make and retain copies.

Pomeroy's response varied. He is quoted as saying, "The first time I saw it I got mad. Then I got scared, and finally I started thinking it was funny. Now I think there is a bigger principle involved."

Pomeroy realized he was the target of a specific investigation solely because he had taken advantage of his constitutional rights. He had cared enough to speak out on an issue which disturbed him, to take sides in a controversy, and to make his name publicly known.

The full impact of the situation took time to develop. Pomeroy went from reaction to reaction. He talked with other friends, and they in turn spoke with their acquaintances. Weeks later, one of the original members of CASE, after hearing of the matter, checked it out. Convinced the story was true, the unnamed individual called Jim Marrs, a reporter on the Fort Worth *Star-Telegram*. Marrs' advice was: "Tell him to come forward and go public with the evidence. When he does that, everything else will fall into line."

More weeks passed. Marrs spoke several times with his contact, but Pomeroy was considering the impact such a disclosure might have on his life in Farmers Branch and in the Dallas area. Then he made a decision. The evidence of the investigation should be made public. But the opportune time was the day prior to the formal AEC hearing on the advisability of TESCO's breaking ground on the reservoir for the new plant. He waited patiently. All the while, rumors of the report circulated through the city's news circles.

On July 30, 1974, Pomeroy and the other CASE members announced there would be a press conference the next day to release information about the alleged investigation. Jim Marrs was informed by telephone, and met with the city editor, who reminded him of an assignment already in progress. Another reporter, Rowland Stiteler, was selected to attend the conference.

Stiteler arrived at the assigned meeting place and was surprised at the lack of newsmen present. Aside from another reporter from a local limited-circulation paper, *The Iconoclast,* the meeting was largely ignored. Two individuals came from the Dallas public-service TV station, and another man represented a radio news station.

In his statement, Pomeroy told of the investigation and the report which had been sent to his superiors. Copies were shown to those present. He stated he felt the dossier indicated he was a Communist or Socialist, but he did not personally know anyone of either persuasion. Further, if he had talked to a Socialist at the city council meeting, he didn't know it. He'd spoken to a number of people. He also said he felt the report had backfired, because he had received no comments from his employer, Continental Airlines, about the matter. He reiterated his stand on nuclear energy, and said he intended to protest the construction permit for the Glen Rose nuclear facility at the upcoming meeting.

His revelations, backed by the copies of the document, failed to make shock waves. *The Iconoclast* ran a story later in the week, and the Fort Worth *Star-Telegram,* on the morning of July 31, printed a two-column account of the press conference.

Then the matter seemed to die. But not for long. The Dallas public-service TV station did a presentation on the problem, and this seemed to wake the rest of the electronic news media. Feature followed feature, and the two other major dailies, the Dallas *Morning Herald* and the *Times,* finally joined the fray.

The reporters on the Fort Worth *Star-Telegram,* moving ahead of the rest, expanded the range of their investigation by locating David Dimick, who refused to talk for publication. The matter was then taken to Colonel Speir, head of the DPS, who promised to "look into it."

Pomeroy's timing was excellent. Just the day before his press conference, he was the subject of a newspaper story on the upcoming hearings. It had proven to be a good platform for him to cite his views, so he'd gone on record as saying the proposed Comanche Peak power plant was unnecessary, too dangerous, expensive to build, and the federal licensing process under which it was to be constructed was "a farce."

His comments met with instant rebuttal, both from the Texas Electric Service Company and the AEC.

Company officials stated the facility was vital to future electrical needs of the area, was safe, and had been well researched to provide minimal environmental impact. The Senior Project Manager for the AEC, C. J. Hale, a native of Fort Worth who was working with the project-licensing division in Bethesda, Maryland, made a quick but limited denial by telephone. Hale said since the meeting was scheduled for the next day, it would be improper for him to respond to specific points, but he insisted the AEC was thorough, professional, and impartial. Both the Commission and the utilities had done extensive research, and the AEC was an advocate of the site because of the many hours of questioning prior to the actual public hearing. He was careful to say the AEC's position was derived without influence by the utility and as a result of its own investigations. He said that by the time a hearing was convened, countless facts had been brought out and discussed with the electric company involved in the construction of the plant. As he put it, it was because "all of the battles and the bloodletting between the AEC and the utility are already resolved by then."

The TESCO officials shared his feelings. "From our point of view, it's no farce," Robert Martin said. "It is comprehensive, it is thorough, it's a well-thought-out plan for licensing. It's tough, they don't fool around."

Martin also said the research for the environmental statement required by the AEC showed the impact would be minimal and the plant should be built. He maintained the study had been done in a professional, scientific manner, fully checked by the AEC. The data had been sent to other agencies of the state and federal governments and "anyone in the United States" could inspect it for accuracy.

"The implication is that we are doing all this as a snow job," Martin went on. "Well, that's not true. We are doing this to design a safe plant."

Pomeroy's rebuttal, printed almost in its entirety, made good press. "They [the studies] are subject to bias from the utilities and they are subject to bias from the AEC standpoint," he maintained. "If there was some flagrant violation, I guarantee you the AEC would catch it. But it might be something that they didn't catch. A lot of things are happening [at currently operating plants] that they didn't expect to happen. For them to arrogantly assume that they know everything that could happen in those sophisticated plants is ridiculous. It is dangerous. And they have done it."

The furor caused by both the confrontation in print between Pomeroy, the AEC, and TESCO, and the revelation of the unusual intelligence investigation and report began to capture public attention. Where, before, CASE had had little influence, there was suddenly new interest in the small organization.

On a state level, the press was making difficulties for the Department of Public Safety. Continued probing finally resulted in a statement from Colonel Speir. He cited a "tip" as being the cause of the investigation, and said little else.

Robert Martin, speaking for TESCO, declared "None of us had anything to do with this."

Finally, details of the incident were shown to Governor Dolph Briscoe and several members of the Texas legislature, resulting in an immediate movement for a subcommittee investigation of the DPS intelligence activities. From the start, it looked as if the motion would pass both houses very quickly.

Colonel Speir maintained the DPS did not investigate people solely because they had spoken out against atomic power. "We have no desire and do not compile data on persons just because of their political beliefs. . . . We have no quarrels with legitimate protest groups." He added, ". . . We are not perfect . . . Somewhere along the line, they are human beings, . . . someone will make a mistake. But this would be extremely rare." He also stated Dimick's sending the report to Continental Airlines was "an error on the part of a young officer who gave it out without permission or authority and in violation of department policy and procedure." Speir issued a public apology to Pomeroy for the mistake and agreed to take disciplinary action against Dimick under advisement.

On August 2, 1974, Governor Dolph Briscoe asked for a copy of the report. Shortly after, the Texas state senate scheduled subcommittee hearings for September.

The Reverend Dwight Brown of the First Unitarian Church of Dallas emerged as a spokesman for his congregation. Incensed by the fact the DPS had listed the church as a leftist support group, he began pondering legal action. Pomeroy announced the filing of a suit against Speir and Dimick, requesting ten thousand dollars in damages. Quickly after, the church also brought action. Pomeroy's suit called for an order prohibiting DPS investigation of politically active individuals.

The hearings were not held until the summer of 1975, and had only one noteworthy incident. Dimick, who was asked by Judge Jack Roberts of the U.S. District Court to produce the identity of the two informants he had relied on in his investigation, at first refused to comply. Then, after a few days, he capitulated and provided the information.

While all of this was going on, the construction permit for the Comanche Peak station was issued and development work for the reservoir commenced. In a way, the publicity generated by the Pomeroy incident served to take some public attention away from the project.

Undocumented reports, along with the Reverend Mr. Brown's testimony, indicate the Department of Public Safety began a very thorough review of its files. Moving with great haste, it had all dossiers on individuals with no criminal record pulled, sorted, and destroyed by shredding machine. Colonel Speir later said all but criminal-activity intelligence reports were being purged from the system.

By late September 1974, the matter had reached its zenith. The state senate subcommittee, chaired by Oscar Mauzy, Democrat from Dallas, convened in special session. One of its first witnesses was David Dimick. He reiterated his story about first having become interested in Pomeroy after a "confidential informant" reported he had seen the pilot passing out literature against a proposed nuclear power plant and had also seen him talking with a known Socialist. Dimick admitted later he had given an "unofficial" copy of his report to the security staff at Continental Airlines. The officer said he had concluded Pomeroy was not inclined to violence against nuclear power stations but "anything's possible."

"Specifically, what did you think it was possible that Mr. Pomeroy might do?" asked Senator Mauzy.

"It had been discussed, I think, and I am not sure, I am not positive, but I believe that the Continental people [wondered] could this guy have the capacity, to be so violently opposed to nuclear energy, to crash his airplane into a nuclear power site," Dimick replied.

It was a dark flight of speculation to imagine the copilot of a 727 seizing control of the huge aircraft and then diving, passengers and all, into a preselected power station in a last, futile, dramatic protest.

"This man," queried Senator Mauzy, "while flying a plane loaded with passengers, might crash it into a nuclear power plant, . . . because he made a peaceful demonstration against the construction of this [plant] before the Dallas city council?" His question brought a murmur of response from the people in the hearing room.

"That was discussed," Dimick replied.

Pomeroy responded to the exchange by calling Dimick's testimony "ludicrous." He pointed out that crashing into a working plant would cause radiation to escape, and "this is exactly the kind of thing we are worried about." He also made a point of the fact Continental Airlines had not removed him from his job as copilot and had in no way troubled him about the report.

Dimick, in wrapping up his testimony, noted he had "never

198

been involved in any unwarranted invasion of anyone's right to protest." But, he concluded, "vigilance is necessary to try to prevent unlawful conduct."

Colonel Speir's statement included the information that since the Pomeroy investigation had been noted, the DPS had tightened its policy to limit intelligence investigations generally to criminal activities. Agents had been forbidden to make probes on their own.

Pomeroy's testimony before the committee, which also included Senators Bill Bracklein of Dallas and Tati Santiesteban of El Paso, indicated he had never experienced mental problems and his family had no medical history of mental aberrations.

"Did the thought ever occur to you," asked Mauzy, "to crash an airplane [into a nuclear power plant site]?"

"No, the other members of the crew would probably take a dim view of it." Pomeroy's response brought laughter from the assembly.

"I, as a passenger, would object," responded Mauzy. The room laughed again.

Pomeroy stated he had not known of the investigation and probably would have continued with his protest even if he had been aware of the probing. Mauzy asked if he would have continued if it had meant the loss of his job.

Pomeroy pondered his answer. "If I thought I was going to lose my job, I would have to think long and hard. I like my job. An airline pilot is pretty useless once he leaves one airline."

Several interesting points came out during the hearings. Dimick testified he was, at the time, justified in making a report on Pomeroy according to then-existing DPS policies. He was not certain if the conversation about the possibility of Pomeroy's flying an airplane into a nuclear power plant had been with the Continental security staff or with an unnamed informant. He also maintained the report, due to the coding used to indicate the reliability of the information, had served to clear Pomeroy. No explanation of how or why the document was then forwarded to the pilot's employer was presented.

Colonel Speir expressed his hope the senators would be careful in their considerations of new laws concerning police intelligence. "To over-restrict the proper use of the intelligence function in law enforcement will not only result in the loss of police effectiveness, but the real loser would be the law-abiding citizens. . . ."

The Reverend Mr. Brown, noting the seriousness of the situation, charged the DPS with having purged their intelligence files by shredding the documents involved. This action, he stated, made it difficult for the committee to determine the extent to which past DPS activities had "unjustifiably invaded rights of citizens in their pursuit of lawful political activity." He also noted a temptation to conclude the DPS leadership "preferred to live with the suspicion of past wrongdoing rather than submit its files to the kind of scrutiny by this committee" which might uncover misdeeds.

Brown concluded with some comments about Dimick. After noting the man was not a college graduate, the minister said it is "unrealistic to expect partially educated agents to comprehend the complexities of political protest and to be able to distinguish between legitimate protest and unlawful protest. We are forced to wonder," he went on, "how many other David Dimicks there are in the DPS, men who lack the qualifications for doing the job they are being asked to do."

The outcome of the hearings was the development of several new statutes to limit the activity of the DPS in selected areas. It is likely if Pomeroy's friend and superior with Continental Airlines had not given his fellow pilot the benefit of the doubt—and a copy of the report—none of the intelligence activities of the DPS would have come to the public's attention. Likewise, the ability to cite a specific incident provided extra impetus to the hearings, making the committee's job of selling its new regulations to the other senators and legislators a far easier one.

Inquiry after inquiry by reporters, including Jim Marrs, about the disciplining of David Dimick brought little in the way of response. Finally, weeks later, it was noted Mr. Dimick had been "transferred" from Intelligence to Narcotics, his $11,616 salary remaining unchanged.

On January 1, 1975, Robert Pomeroy left Dallas, Texas, and moved to California.

The matter was ended. But there are still several questions remaining. Was Dimick following DPS policy in 1974? Did that policy allow an investigator to focus on an individual for making public statements against nuclear power? Was it DPS policy for reports, supposedly confidential, to be forwarded to protestors' employers?

No one will ever know the full details. The time in which it happened, 1974, was a strange period in America. Many things previously considered to be acceptable behavior for law-enforcement agencies came under fire. The nation was in a turmoil of debate over questions of illegal or quasi-legal activities of the Central Intelligence Agency and the Federal Bureau of Investigation. Civil rights was a concept, rather than a working reality, to many police officers. The feeling that laws were protecting the criminal by tying the hands of those charged with their protection was rampant. And to some degree, it was true.

The changing times called for new methods—and an end to a number of old ones.

Colonel Wilson E. Speir is a capable man, charged with a tough job. It was his duty to bring the DPS through this period of revision and into a newer era of law enforcement. In many ways he succeeded. David Dimick, who somehow emerges as the villain of the piece, acted in a fashion he considered to be in keeping with what was then DPS policy. There seems to be an undertone of surprise in his testimony, and it is easy to see why he might have been startled by the barrage of publicity and subsequent notoriety.

The job of a police officer is not an easy one at any level. Decisions which affect the lives of individuals must be made on a daily basis. Some of those are going to be wrong. Perhaps, today, after the turmoil and readjustments of the early 1970s, law-enforcement officers will have more respect for the individual's rights and privacy.

But the nuclear industry poses a new problem. Everyone recognizes the atomic power station as a dangerous device. There is strong emotion directed against many such facilities. Law-enforcement agencies, on both a local and state level, will be especially attuned to problems at the plant. And this will extend to problem-makers.

As the nuclear industry builds a force capable of policing and protecting the units, more and more interface will occur with the lawfully appointed police. Will this increased familiarity lead

201

to new levels of cooperation? The answer is probably affirmative. Will this closer relationship between private and public forces bring on new excesses? The potential is there. Responsible management of both groups, the police and the utilities' security staffs, will prevent this. It is not a problem, yet. But it is something we all should be aware of—and watch.

14

■   The advent of Sputnik marked the start of a whole new era of technology. After the first few fledgling attempts, the masters of space began to look for ways to increase the efficiency of their orbital vehicles.

Additional electrical power provided one gateway to the improvement of this new astral breed, and the surest, easiest way to gain large amounts of electricity was by the development of lightly shielded nuclear reactors for use in the instrument packages.

In a way, then, the initiator of the space race was also the instigator of a new problem—that of orbital masses of highly radioactive materials.

No orbit is permanent. All, because of the delicate balance between gravity and velocity, decay, sending the suspended object closer and closer to the point when equilibrium fails and the vehicle falls slowly, then faster and faster on a re-entry course to destruction.

The return of one of these nuclear power piles is not like the homecoming of an ordinary space vehicle. As it disintegrates, it leaves a high-altitude trail of burning dust behind. Each mote is as radioactive as its partner, and each will finally settle to some point on old mother earth.

Other, larger component pieces of the vehicle will also survive the intense heat of re-entry, allowing an estimated 5 to 10 percent of the total satellite to arrive back on our planet in various-sized hunks.

If a portion of the surviving material is from the reactor unit, there will be a highly radioactive piece of debris traveling at an enormous speed toward its specific point of impact.

If this predestined, almost-preordained spot is a center of population—New York City, for instance—an inestimable amount of damage will result.

Although it is not generally known—or wasn't at least until the coming of Cosmos 954, which gained headline prominence in newspapers throughout the world—several satellites have made the re-entry journey complete with their nuclear hearts.

Cosmos 954, however, is the one to remember. As a story it has almost everything: suspense, international intrigue, special operations with code names, flights in search of vital intelligence materials, treks across the hostile, frozen tundra, and more.

A look at the behind-the-scenes story will reveal many things—and will possibly remind us, the next time we gaze upward on a star-bright night, of a new kind of proliferation. One which shows every sign of becoming another rather nasty facet of our already complex atomic age.

■ Silence. Blackness. Even though the 46-foot-long Cosmos 954 was moving at over a thousand miles an hour, there was no sound, no rush of air. On board, a specially designed nuclear reactor, fueled by about a hundred pounds of uranium 235, generated more than enough kilowatts to provide power for the sophisticated electronic equipment the Soviets had painstakingly installed.

Launched on September 18, 1977, into a 150-mile orbit around the earth, the 8,000-pound vehicle was ill-fated from the first. Ground tracking stations began to notice uneasy signs in its flight stability and responsiveness from the fifth orbit.

No official information has ever been released on what the purpose of this particular satellite was, but that is hardly a departure from the norm. Over a thousand functioning earth-orbiting man-made moons are in place, performing a variety of tasks ranging from television transmission for civilian reception, to military spying. In addition to weather and navigation control, designs have been launched to scan the seas to keep track of shipping, to guide ballistic missiles to targets, to monitor launchings of other rockets, and to perform a vast number of intelligence functions.

Speculation on the true role of Cosmos 954 has ranged from reports indicating the craft was equipped with a unique radar system to monitor U.S. missile-firing submarines, to a science-fiction-style scenario in which it was designed to be a "killer," using a laser beam or a kind of death ray to destroy other space vehicles or possibly even ground targets. (The Soviets do possess killer satellites, but they are simple, surface-directable proximity bombs, not flying ray guns. Some qualified reports, however, indicate that the Russians are experimenting with various plasma devices for use in this area. They are just not operational at this time.)

All the various ideas make sense and are in keeping with the current state of the art in weaponry and surveillance. And all gain credence because of the size of the atomic pile on board. The relatively vast amounts of electricity produced by the one-meter-square reactor from 50 kilos of fissionable uranium must have had some end use. And since the original launch site was in the heart of a Soviet military complex, the satellite obviously had an intelligence role aimed at the western nations. The United States began a careful study of the Cosmos shortly after its attainment of orbit. A visual analysis of its exterior was made from photos obtained by both ground and space installations, and several arm's-length tests were undertaken through the use of electromagnetic spectrum analysis, radio reflection, and other advanced methods.

In some ways the satellite was familiar to our scientists. The first of the Soviet ocean surveillance series was Cosmos 198, launched in 1966. Since then, sixteen more had been fired into orbit. All apparently were intended for a very specific reconnaissance job. Each behaved in an unusual fashion.

After attaining an orbital plane about 260 kilometers high, with an inclination to the equator of 65 degrees, they would stabilize. Then, a few days later, at a command from the ground control, on-board booster rockets would fire and shift the vehicle outward to a distance of 900 km. The very low initial orbit for this series is a result of a limited resolving power of the spy equipment on board. One revolution takes 89.65 minutes, which is dangerously close to the critical time of 87 minutes, at which point the object loses stability and will fall back to earth. The move outward to the higher parking orbit was designed to keep the on-board atomic reactor from plunging back into the atmosphere.

The first six Cosmos shots were experimental. But by 1973, the device was considered perfect enough for military use, and since then ten more have been lifted up. All but two worked in pairs to provide ground stations with the speed and the direction of moving ships. It appears they are used to monitor special events, rather than for continuous surveillance. In 1974, a set was orbited to keep track of the NATO naval exercises held in May. Our familiarity with the series was limited, so the opportunity to study another one while it was studying us received top attention.

The newly launched satellite was tracked in a small green room buried over half a mile deep in the solid granite of Cheyenne Mountain. Located in the state of Colorado, this unique installation is the functioning heart and headquarters of the North American Air Defense Command (NORAD). Atomic bombproof, and protected by one of the most elaborate security systems in history, this nerve center processes phenomenal amounts of information each hour of the day, ordering the data into meaningful displays and readouts so the officers, scientists, and technicians can assist in regulating the retaliatory strike force of the United States.

One of its least-known duties is the maintenance of the "junk catalogue" of orbiting objects. Starting in October 1957, the nations of the world began sending up various payloads and suspending them in the sky by carefully matching their velocities and altitudes. Since the beginning of the space age, about 10,500 items have been orbited. Around 6,000 have dropped back into our dense atmosphere—which leaves 4,500 still circling. NORAD is charged with maintaining watch on each and every part of this expensive space junkyard.

The need for such careful surveillance was not immediately apparent in the 1950s. Several things were shot off before the NORAD logbook began. By February 1978, the latest entry in the catalogue was item number 9,644. At the same time, 4,546 items were under watch and still in full orbit. They ranged in size from less than a foot square (a Hasselblad camera which slipped away from a space-walking astronaut in the early 1960s) to the massive 84 ton Skylab scheduled to fall in late 1979 or 1980.

Accurate tracking and position reports, along with information about times and places of anticipated re-entry, are furnished at U.S. expense to the Soviets and the rest of the world, in an effort to prevent a nervous radar monitor from setting off World War III by mistaking the blips on a screen caused by the return of some junk for the final homing path of a multiple warhead intercontinental ballistic missile. The number of times NORAD's special service has prevented a serious misunderstanding is unknown, but it has saved us on more than one occasion.

206

Blue-uniformed technicians and officers analyzed the data on Cosmos 954 for forty-four days, expecting the Soviets to fire the on-board rockets and move it into higher orbit. During this time, the craft made several small random maneuvers. On November 1, these stopped. Something had gone wrong. The vehicle showed every indication of returning to earth. Where the multi-ton radioactive hulk might fall was open to a number of guesses, but there was little question it was coming down.

Official notice was passed on a routine basis to the National Security Council (NSC), and a message came back to NORAD requesting a tighter fix on the date. By mid-December the situation had progressed to the point at which re-entry was certain and a number of possible sites were listed. Even at this early stage, a North American impact point seemed likely.

The Soviets, aware of their failure to move the vehicle into its parking orbit, tried other techniques. Utilizing a built-in remote-control rocket engine, they attempted to split the satellite into three separate portions. Two, not containing the atomic reactor and therefore only slightly contaminated with radioactivity, would be allowed to fall back. The third, with the pile inside, would be shot outward, where it would remain for a thousand years. This appeared safe enough to Soviet scientists because by the year 3,000 mankind would either be able to retrieve the hulk and move it into a collision course with the sun, or would be decimated to the point at which one more radioactive device falling from the sky would make little or no difference.

Two separate attempts were made by the Russian technicians, but in both cases the rocket refused to function. It is likely they managed to disconnect the parts of the soaring satellite but they were unable to achieve a successful reorbiting of the pile.

Alarmed by the problem, the Soviet team arrived at the same probable re-entry sites listed by NORAD. Knowing for certain we had complete knowledge of the coming event, they did not make the fact of re-entry public. They also did not bother to mention it to the U.S. or Canada.

The NSC officials had kept themselves current on the matter. When NORAD estimates improved in late December, and the North American continent had become the number-one impact site, President Carter's National Security Advisor, Zbigniew

Brzezinski, after talking with the President, set up an appointment with the Soviet Ambassador, Anatoly Dobrynin.

Dobrynin's background as an aeronautical engineer gave him a good basis for understanding the erratic nature of the re-entry path and the potential seriousness of the matter.

During the meeting between the two on January 12, several impact points were contemplated based on the latest data. According to Brzezinski, there was a "serious hazard to the public" if Cosmos 954 were to fall into a densely populated area. While the touchdown predictions did not indicate it, one orbit occurring in the final unstable minutes could cause it to strike near New York City.

The main subject of conversation, however, was the nature of the radioactive material on board and the probability of its attaining critical mass from either heat generated by re-entry into the atmosphere or impact upon landing. Brzezinski sought information from the Soviets on points of design as well as their safety features, and after the meeting he said Dobrynin's replies had been "somewhat reassuring but not fully satisfactory."

Projections based on the amount of atomic material thought to be on board the spacecraft indicated it could potentially have the explosive power of about five times the force of the bomb dropped on Hiroshima. The chances of an explosion, however, were slim by everyone's estimate. Nonetheless, the radioactivity which might be released was a matter of vital concern—especially because of the possibility the object might fall into a large city.

A second meeting between Brzezinski and Dobrynin was held on January 17, and at least two telephone calls were exchanged in the next few days.

The White House staff decided to prod the Soviets for more information and to alert the leaders of selected countries which could help in the tracking program. Later, information was extended to all our North Atlantic Treaty Organization (NATO) allies and to Japan, Australia, and New Zealand. No public announcement was made because, in the words of a spokesman, "We were trying to head off a re-creation of Mercury Theatre." The reference was to the 1938 Orson Welles-produced radio drama, "War of the Worlds," which caused widespread panic among its listening audience.

During this same time, other related events were taking

place. Intelligence teams from the CIA, the NSC, and the Air Force began playing with the idea of trying to recover Cosmos 954 to see what might be learned from it. The more they developed the idea, the more inviting it became. Later, after its re-entry, there was speculation the U.S. might have had some role in the downing of the erratic satellite, but there is no available evidence even pointing to such an act. It is possible, but not too probable.

NORAD's reports still indicated the impact point would be somewhere in North America. It was pinpointed even more closely to some part of Northern Canada, but due to a skipping effect which occurs as the re-entering vehicle first touches our dense atmosphere, no entirely accurate touchdown point could be given. Unless there was a sudden change in the stability of the satellite, though, it seemed probable it would not enter U.S. territory.

On January 19, in reply to diplomatic pressures, Dobrynin contacted Brzezinski and informed him of the nature of the radioactive material on board the Cosmos, revealing it to be uranium 235, in about a 100–110-pound quantity. The Soviet ambassador made it very clear the core was designed to prevent its going critical during descent or impact. He also gave all possible assurances it would not explode, citing the excellence of Russian engineering and the stability of the type of nuclear power plant on board. "He wanted it understood that the Russians were not orbiting a nuclear bomb," one of the State Department officials was later quoted as saying. The use of the word "bomb" is interesting, because this possibility must have been given consideration.

Even though there are treaties and agreements concerning the orbiting of nuclear weapons, there is no formal or forceful policing procedure to see to it the participating nations are in compliance with their words. Brzezinski wasted no time after the meeting. He issued a National Security Council directive alerting NASA, the CIA, and the Departments of Defense and State.

The response was quick. Special Air Force teams trained in radioactive detection and decontamination were placed on standby so they could be flown to any given spot in a matter of hours. U-2 aircraft were fitted with high-altitude sensors, and portable detectors were readied for use in low-flying ground reconnaissance transports.

By the third week in January, NORAD placed the re-entry over the Queen Charlotte Islands off the coast of British Columbia, the Canadian province bordering on the Pacific Ocean. The impact date was estimated to be January 24. All branches of intelligence gave their plans a final once-over, and the teams of specialists drew issues of arctic clothing to protect them from the anticipated −20 degree temperatures common to the Canadian tundra. On the evening of January 23, Operation Morninglight was like a cocked pistol. The first sighting of the flaming wreckage would pull the trigger.

It was cold in the early dawn of the 24th, and in the far northern area near the Great Slave Lake there was a lingering semitwilight. The habitual gray cloud cover was high and the fierceness of the freezing cold prevented any but the most necessary excursion from the row of houses lining the main street of Yellowknife. The town, with a population of about ten thousand, is located nearly a thousand miles north of the border between Montana and Canada. It lies in the heart of the Northwest Territories, the immense, chill Canadian outback or wilderness. Yellowknife exists because of the gold mined in the area, but it is a settlement with two distinct personalities. In the summer, when the days are long and light, the people move about outside. Since mining can be conducted only while the ground is soft enough to work, labor starts early and ends late. In the winter, intense cold, with temperatures falling to more than 40 degrees below zero, forces the residents inside. Blinding blizzards of snow blow through on a regular basis, adding to the problems of the settlers. Only the rugged remain.

Marie Ruman, a pleasant woman, was standing in the office building where she worked, looking out at the gray, dim overcast when something caught her eye. She glanced away, then back again, giving a little cry of surprise. "I looked through the window and saw this object coming toward me," she later said. "The main part was like a bright fluorescent light. There were lots of small parts trailing behind it. The pieces were bigger than shooting stars and each had a long bright tail. None of it made a sound." Startled, she called others. Cosmos 954 had returned to earth.

Dale McLeod of the Royal Canadian Mounted Police steered his patrol vehicle carefully on the frozen rough road. The country seemed quieter than usual and the bleakness of the snow-covered, pine-tree-spotted landscape presented a serene presence to the two men driving through it. Suddenly a fiery trail shot across the drab sky and McLeod slammed on his brakes. The brightness of the fireball tingled his eyes as he watched the mysterious sight. "It was like something out of Superman. You know—it's a bird, it's a plane, no, it's . . . it's . . . it's whatever." Astounded, he used his radio as he and his patrol partner watched the glowing object move rapidly toward the horizon.

One hundred twenty-five miles south of Yellowknife, in Hay River, another Mountie saw the light. Corporal Phil Pitts had gone up onto the roof of one of the RCMP Detachment's buildings in the compound and spotted a "bright white and incandescent" glowing object tracing a path through the dim sky. He immediately reported the incident, and was informed it probably was the Cosmos re-entry. Later, after learning of the high levels of radioactivity associated with the returning space vehicle, he is quoted as saying, "My gosh! I was standing on the roof watching it go by. Maybe I'm sterile."

During the night of January 23, the emergency teams remained on full stand-by alert. They were ready to move. NORAD had set up a special communications link and passed the word to the White House as their computers refined the rapidly developing data. Later in the evening they gave a landing time of about 6:53 A.M. on the morning of January 24. This was duly relayed to all concerned parties.

Brzezinski was on top of the situation in the early hours of the 24th, personally telephoning President Carter, whom he woke, with the news of the actual re-entry. The NORAD tracking system, aided by other ground station radar, had followed Cosmos during its descent until it finally lost sufficient altitude and fell off the screens. Its last known height and position were carefully noted, as were its ballistic characteristics.

The general impact area was near the Great Slave Lake, and since radar had indicated some fragmentation during the later stages of the fall, a corridor about 1,500 miles long was plotted. All the debris would, or should, be within this long, narrow zone.

After telling the President, Brzezinski called Ambassador Dobrynin to tell him Cosmos 954 was down. Then, at 7:15 A.M., EST, he called Canadian Prime Minister Pierre Trudeau with the position estimates. Twenty-two minutes after the final fix from

NORAD, the leaders of the concerned nations knew of the averted disaster. Trudeau, who had been briefed earlier in the project, had known since the previous weekend the object would fall inside Canadian boundaries. He too had agreed not to alert the public to the potential menace.

Once the final impact zone was delineated, Operation Morninglight shot into action. A KC-135 and a U-2 high-altitude reconnaissance aircraft were given sampling locations and they commenced a long aerial patrol along the satellite's path. Checking for various kinds of radiation, these vanguard planes made several sorties but returned each time with negative results. There were only limited signs of upper-level contamination. At the same time, over a hundred U.S. scientists, technicians, and soldiers embarked into the freezing winter wilderness aboard four C-130 Hercules cargo planes. A 22-man Canadian nuclear-accident response team, equipped with various kinds of radiation sensors, was dispatched to Yellowknife. They patrolled the streets and the surrounding countryside, moving through deep snow, but found no unusual indications of radioactivity.

A special 44-man United States Air Force task force made up of very carefully trained and selected specialists was dispatched to the Morninglight control center in Edmonton from Andrews and Nellis Air Force bases.

Our intelligence teams were also on the move. There was a good chance the satellite's remains could be found by seeking out the puddle of radioactivity which had to be formed around any remaining pieces of the reactor segment of the vehicle. Every chunk which could be located would serve the intelligence engineering staff well. From even small fragments they would be able to deduce the state of Soviet reactor design, something about the electronics package the unit carried, the latest thinking in Russian metallurgy, and hundreds of other technical facts.

Within hours of the final touch down, men wearing "man from Mars suits" which covered the head, face, hands, and feet, were in Canada to search out the debris.

Around-the-clock flight plans were instituted for the large cargo planes, and as soon as the engineers had installed the gamma-ray spectrometry equipment which would serve as their bird dog by pointing out the hot spots as the aircraft flew by, they were off.

Pilots, crews, scientists, United States Air Force nuclear specialists, engineers, and others all gathered in the Edmonton headquarters, awaiting the need for their skills. Additional increases in service manpower to provide meals and quarters for the new arrivals almost doubled the total number of people involved.

The flights were long and tiring. An eyewitness aboard special Flight 6763 provided details of the mission.

It was midmorning by the time the ground crew had refueled and checked the Hercules C-130 and released it to pilot John Oliver. He, along with Serge Cothe, his copilot, did a careful preflight. Both men knew they would be flying at low altitudes over some of the most bleak and desolate country in the world. The strange terrain was an endless field of snow and ice, broken only by an occasional outcropping of black rock or a scattered grove of wind-blown pine trees. Even with the latest radio and ground-search techniques, a crashed aircraft was very hard to locate there—and downed crews had only a limited amount of time because survival in the intense cold was difficult.

Satisfied with his walk-around inspection, the pilot clambered aboard the plane where the technicians had completed the checkout of the two 1,300-pound spectrometers. The large, square, off-green instruments would be the heart of the search system. Oliver spoke briefly with Peter Holman, the geophysicist who would evaluate the readings from the units, then moved into the cockpit, where he checked his maps. Working with Cothe, he reviewed their navigation preparations. Then, settling back, the two men went through the elaborate before-takeoff checklist. Once airborne, they flew directly to the assigned search zone to begin a seemingly endless crisscrossing of the icy terrain.

Holding the flying altitude down to under a few hundred feet gave the green boxes in the belly of the ship a better opportunity to pick up any signs of radiation. But because of the low altitude, each search leg was narrowed. Oliver and Cothe took turns at the controls, each man concentrating on holding a constant airspeed, altitude, and course. Occasionally their radio would crackle into life and they would exchange brief words with a ground station or another aircraft. Only a cup of coffee from a battered thermos broke the monotony.

On the seventeenth pass the instruments registered some sign of radioactivity on the frozen tundra below. Notes were made

213

on the maps and an electronic fix of the plane's position was attempted. Asked later if the indicators on the green boxes had shown a piece of the fallen Soviet spaceship, Holman shrugged and responded, "I don't know, but it's worth looking at."

The search continued while other crews, responsible for taking the closer look indicated by Holman, moved into action. These specialists, trained to deal with radioactive materials and equipped with smaller radiation detectors, were maintained on standby to be dispatched for on-site ground examination. Airlifted by helicopter, they could cover a number of possible spots in a given day. Their routine was arduous.

Given the map location of the suspected area, along with any electronic navigational information, fixes, and visual marks reported by the C-130 survey crews, the men would land in the general vicinity of the position established by the gamma-ray spectrometers. Using this point as a center, they would then conduct a low altitude hunt, following a spiral pattern, to see if any equipment in the helicopter would give them a closer indication. If so, they landed and would begin an even finer search, utilizing hand-held instruments and making their way across the hostile wastes on foot. They would stay in the selected area until every square inch had been covered by both visual inspection and radiation counter. If pieces of the Cosmos were present, they would find them.

The search effort continued for several days. Then, an unusual series of events occurred.

Barney Danson, the Canadian Defence Minister, publicly indicated the field efforts had turned up "either a piece of debris or the greatest uranium mine in the world."

His statement resulted in increased interest in the project by the press and his announcement was carried on American television.

The next day, though, official Canadian spokesmen indicated the Danson report had been a false alarm. In explaining the discrepancy, some vague "equipment failure" was cited as the cause.

Apparently this caught the American contingent involved in Morninglight off guard. The crew chief in charge of the green boxes was a highly respected U.S.A.F. nuclear physicist. His

214

reputation for efficiency was impeccable and the idea there had been an equipment malfunction which was wrongly read as a ground source of radioactivity didn't fit with the team's past performance. After twenty-four hours, the United States replied in a terse statement indicating the instrumentation had performed as expected and "Something hit the ground that was radioactive." William Nelson, a U.S. Department of Energy scientist, came forward with a statement, saying, in part, "Pieces of that satellite *did* impact the earth."

What caused these strange comments is still not clear. The Soviets, while admitting no liability, had on several occasions asked the Canadians if they could join in the hunt so as to be allowed to clean up their own mess. Washington, naturally, was against Russian assistance as intelligence experts wanted to examine all the pieces which could be retrieved. And both the United States and Canadian governments were concerned about leaving radioactive debris scattered about the countryside. In the end, the United States allowed the Canadians to deal with the Soviets—which they effectively did by ignoring their requests to participate and subsequently billing them for about a million dollars in search costs.

The admission of success, followed by the denial and a final acknowledgement, might be related in some way to the political maneuvering which went on during this period. Or it may be only the visible aspect of an entirely different manipulation. It also could have been the result of an honest series of mistakes caused by a confusion of communications during a tense and hurried time.

In any case, even though there were signs from the overflight portion of Morninglight, as of January 27 no one had actually seen a single component of the crashed Cosmos. The search area encompassed 15,000 square miles of wilderness in a long northeasterly corridor from Canada's Pacific Coast to the Great Slave Lake. In the grip of winter, daily cold of −30 to −40 degrees was the norm. And with the chill factor, caused by the unceasing wind, the temperature would be experienced as 80 or 90 degrees below zero.

This land has defeated even seasoned outdoorsmen. John Hornby, leader of an expedition in 1926, had attempted to cross the Northwest Territory from the Yukon to Hudson Bay. He tried to winter in the region but it proved too much for him. He died there.

The story of Operation Morninglight now takes a strange turn. An expedition put together by a young Canadian and five American companions set out in the spring of 1977 to retrace the steps of the original Hornby group. Finding themselves well into the wilderness by the first snowfall, the team decided to winter in the area. For a headquarters, they selected a fifty-year-old log cabin near an almost unpopulated settlement called Wardens Grove.

It seems odd a group composed of five Americans and a single Canadian was in the area of the Cosmos disintegration path. And while it is possible there is more to this than happenstance, the parents of the Canadian, Robert Common, knew of his plans for at least a year before the Soviets say they launched their sky spy.

From their base camp, the six young adventurers planned to carry out a number of activities. The Canadian Government had given them a winter wildlife study and they were also charged with conducting a number of different meteorological surveys. In addition, for their own satisfaction, they wanted to travel across country about twenty miles to the site of Hornby's last camp.

Since some of the team would be needed to man the base, they divided into three sets of two. John Mordhorst, of Rock Island, Illinois, accompanied by Mike Mobley, of Mesa, Arizona, set out by dogsled to the Hornby site. They made good time, and after a night out, completed their visit and started back to the log cabin at Wardens Grove.

About 3:00 P.M. on January 28, the two men were moving along the frozen mass of the Thelon River when something caught their eye. It would have to have been something pretty unusual because with the temperature in the −40-degree range their attention was totally on their own cross-country path. They did not know of the satellite's fall. Interested, they changed their course. In a matter of minutes the dog team had drawn up alongside a wide "crater about ten to twelve inches deep in the snow." They discussed the unusual array before them and one of the two ran his mittened hand over a metal spar jutting out of the white

216

ground. Describing what they found, they later said, "there were prongs, struts, a tripod shape with the apex in the ice." Curiosity satisfied, the two refigured their direction to compensate for the brief detour and returned to the relative warmth and security of the cabin.

They arrived to find their friends excitedly discussing the re-entry, and a few minutes of conversation convinced them they had stumbled onto pieces of the wreckage. After discussing several possible courses of action, they radioed the authorities.

Hours later, the two who had actually seen the debris were the subject of an extensive medical examination to determine the degree of contamination they had suffered. A special team arrived by Chinook helicopter from Baker Lake consisting of thirteen Canadians led by Lieutenant Colonel Donald Davidson. Backed by a U.S. crew headed by Paul Murda, they airlifted the two explorers out, to be followed by their friends and even the dogs, so whole body counts could be performed to assay their condition. Fortunately, neither Mordhorst nor Mobley had severe dosages. Their exposure was about equal to two chest X rays. The other men in the group, as well as the dogs, fared better still. Concerned about the possibility of radioactive beryllium, which is hazardous because it can be picked up through skin contact and then swallowed or inhaled, the nuclear experts earmarked a piece of the metal at the site for analysis. After exhaustive physicals, all six men were returned to their Wardens Grove camp within a few days with clean health reports.

Davidson, with his Canadian team and part of the American contingent, followed the directions given them by Mordhorst and Mobley. They flew their helicopter to an outcropping of black rock near the site, then made their way through the snow to the spot.

According to Lieutenant Colonal Davidson's comment at the scene, "Something has really gone through that ice at high speed. This is all that's left sticking out."

Their instruments indicated the materials still showing were radioactive, but not sufficiently so as to be classified as a part of the reactor unit. The two teams set up camp and began working on ways to remove the broken pieces while maintaining a safe level of exposure.

The Thelon River discovery was only the first of the finds. Two days later another ground party discovered debris on the snow-covered ice of the Great Slave Lake. The hunks were smaller, on first examination, but thinking there might be more under the water, the team established a field outpost at Fort Reliance.

The Slave Lake material was far more dangerous than anything previously found. One chunk, ten inches by three inches by about a half inch in thickness, gave off more radiation in a single hour than forty times the allowable limit for an employee in an atomic power station to acquire in a year. To move in close enough to work, the men had to stay behind a 1,600-pound lead shield, and all pieces were handled by long tongs. The effort was further hampered by a 35-mile-per-hour wind which took the chill factor down to more than 100 degrees below zero.

Reports since the January 28 discovery have become fewer and fewer. As this is written, crews are still in the impact zone. And some thought has been given to establishing a summer project to probe the river and lake bed after the ice has melted. Sources also say our intelligence experts are very pleased with what they have to date—which indicates they found enough to form some idea about the sophistication and operation of the satellite.

Cosmos 954 was not the first orbital object to crash to earth carrying a nuclear-fueled reactor. To date, there have been at least six such incidents. None of them gained much publicity because they did not threaten population centers. Three of the wayward devices were ours and three were of Soviet origin.

The U.S. vehicles include a Nimbus weather satellite launched from Vandenberg AFB in 1968, which plunged back into the waters of the Pacific after failing to obtain orbit; a 1964 incident when a Navsat navigational unit came down over the island of Madagascar; and the unused moon-landing module from the ill-fated Apollo 13 mission.

In 1969, two Soviet unmanned moon-mission landers fell back, and another U.S.S.R. satellite was lost in 1973 somewhere in the Sea of Japan.

Currently, according to latest reports, we have nine vehicles in orbit carrying some form of atomic pile. Eight are of the plutonium variety; one is a uranium model. The United States is active in the area of nuclear-powered space probes with the

218

so-called SNAP program. Systems for Nuclear Auxiliary Power is a well-funded, far-ranging engineering development effort to provide us with the capability of 10-kilowatt-energy generators for advanced satellites.

Of the three U.S. power packages which have fallen back to earth, one, from the Nimbus weather satellite, was recovered by a small submarine. The Apollo 13 unit was jettisoned deliberately and fell into the South Pacific; there have not been any signs of radioactive leakage. No trace of the Navsat has ever been found.

We have little or no information on the Soviet mishaps, except, of course, Cosmos 954. Reports vary, but it appears the U.S.S.R. still has about eleven other reactor models flying. Simple addition of both U.S. and Russian vehicles indicates about half a ton of enriched uranium and a hundred or more pounds of plutonium are circling overhead—not to mention the few kilograms left on the surface of the moon during the Apollo programs.

There is a major difference in the power requirements needed by the technologies of the two countries. In the United States, we have used, in all but one case, a 70-watt nuclear battery called a radioisotope thermoelectric generator. The U.S.S.R., on the other hand, flies a true reactor, capable of generating between 10 and 100 kilowatts.

The lower energy levels aboard the American ships stem from our lead in the development of infrared and multispectral scanning devices to provide information. Russia has depended upon more and more sophisticated radar, mostly to track our ships and subs, and thus has a higher power requirement. Our SNAP program indicates we feel the need for more advancement in the area of on-board satellite power systems, and in fact the only uranium pile we currently are flying is SNAP 10A, launched in 1965 as an experimental vehicle to gain practical experience. This in turn may mean we are interested in improved radar as an addition or backup to our present scanners. Or it may indicate we too are active in the design of field-effect or ray-type weapons.

The Cosmos incident is important because it calls attention to another aspect of proliferation. While there is little chance of either nation's satellites exploding like a bomb, the possibility exists that a large-population area might be exposed to high levels of radiation. In any case, the upper atmosphere could be contaminated over a wide area. There is a strong probability most of this material would reach the earth eventually.

Recent events show a hesitation on the part of the two nations to halt their space-related atomic activities. The United States, in fact, is scheduled to launch a new pile inside the 1982 firing of the Jupiter Probe.

Several "cures" for radioactive litter have been brought forward for consideration. The re-usable orbiting space shuttle program might be used to pick up a dangerous object before it drops back into the atmosphere, then carry it out to a new, higher plane of rotation. Or, a special killer satellite could be sent aloft and guided to a rendezvous with a wayward object. Once in position, it could destroy the reactor unit before it falls. Both suggestions are workable. But it will be interesting to see if either is ever carried out. In the meantime, Canada will strive to collect the cost of the satellite search from the Soviets, and the Tundra Silkscreen Company of Yellowknife near the Great Slave Lake will make a buck by selling commemorative Cosmos T-shirts.

It is a rather fitting end to a story of potential modern peril because it is so typical of our cavalier attitudes.

·15·

■ On a clear, bright, dry day in 1976, a huge 18-wheeler tank truck carrying a cargo of liquid ammonia charged into an elevated curve over the intersection of Loop 610 and the Southwest Freeway near downtown Houston, Texas.

A second later, the battered remains of the shiny stainless-steel vehicle lay in a flaming, smashed mess on the roadway twenty feet below. It had struck the guardrail, vaulted over, and dropped into the oncoming lanes of vehicles. The violent force of the accident killed the driver as several automobiles collided before the flow of traffic could respond to the sudden blocking of their pathways.

But that was only the beginning. As the cars and trucks stopped, the freeway became a bumper-to-bumper parking lot. A mile back from the accident, well out of sight of the wreckage, drivers came to a puzzled halt as the string of vehicles stretched endlessly over the crest of one hill, then down, and up another.

Those equipped with citizen-band radios called ahead and queried cars on the opposite side to find the cause of the congestion. The answers they got back almost started a panic.

The truck, leaking liquid ammonia from the broken, crushed tank, was the epicenter of a deadly circle. As the chemical warmed, it vaporized into a highly pungent but totally invisible gas. Heavier than air, it clung low to the ground, flowing like an unseen, noiseless specter.

The people outside their cars, surveying the damage to a fender or gaping at the ripped, smoldering truck, were the first to succumb. A sudden wash of burning, stinging air swept over them. First, their eyes reddened. Then, as they inhaled it, they were gasping for breath, lungs burning from the irritation. One man shouted, "Ammonia!" Another called, "Run!" as he turned from the truck and started a staggering trot away from the source of the

221

trouble. He made about 20 feet before falling to the ground in a shapeless heap. "My God!" another called. He was more successful, and managed to run back through the lanes of parked cars.

A man sitting in a pickup equipped with a C.B. recognized the scene. Experienced with chemicals, he realized the overturned truck was spewing out thousands of pounds of toxic liquid. He opened his door and got the first faint, faraway whiff of sharpness. Grabbing his microphone, he relayed his impression of the problem over the channel reserved for emergency messages. Then he shut off his engine, jumped down onto the concrete roadway between the lanes of traffic, and began warning the other drivers to evacuate.

Isolated in their parked automobiles, they looked at the shouting man with disdain and amusement. Inside, air conditioners humming and windows rolled up, more than one person silently agreed with the poor fellow's agitation. Freeways were becoming impossible.

The gas, still clinging to the ground, rolled silently forward. Sucked into the air intakes of a number of vehicles, it brought immediate response. Doors flew open and people dashed out, staggering, wiping their red, crying eyes while gasping for breath.

As the vapor spread, more drivers and passengers bolted from their cars. Some made it to the verge of the pavement, then fell. Others, more hardy or less exposed, succeeded in attaining enough distance from the creeping poison to regain their breath.

People died. Others were injured. Traffic was snarled for several hours as the cleanup crews working with emergency vehicles brought things slowly back to normal.

This was a non-nuclear accident which involved ordinary people. It's a close parallel to one of the possible scenarios which might develop in case of a major problem at an atomic facility. Cars trapped on the freeway, people in nearby homes subjected to radioactivity due to contamination of the environment, and a lasting effect on the countryside. (In the case of the ammonia release, all plant life within a half mile radius of the accident turned brown and died.)

What can an average member of our society do to protect himself and his family? How can a person contend with such an incident?

These are important questions to ask, but the answers are

nebulous. Because inured as we are to the possibility of catastrophe in our everyday lives, we give little thought and less action to such events. There are several things which any of us might do—if we are so inclined.

First, we can examine the planning done for us by others.

The emergency plan is a vital part of the safety precautions of every nuclear facility. But there are two problems with such a predetermined program. Individuals charged with its implementation may hesitate to act soon enough to achieve maximum benefit, due to the large numbers of people who will be involved and the trouble which will be caused.

Second, the very fact of the plan itself is a dilemma to the operators, because their public stance must be one of safety. The existence of an emergency program does not seem to speak of secure operating conditions.

Civilian flying schools have a similar difficulty. Such schools make money from teaching students to fly. The more people who are not afraid of being in control of an aircraft, the more students. One way to quench fears is to liken flying to driving and to maintain its complete safety. It is not too good for business to have an instructor pilot constantly harp on emergencies and dead-stick landings. So this aspect of training is confined to the minimum possible instruction time. Military flying schools, on the other hand, demand high proficiency in these techniques. A student spends a good part of training on the grimmer aspects of aircraft malfunction and powerless landings.

The nuclear utility and other atomic installations are in a similar position. If their processes are safe, then a backup emergency program is superfluous. If, however, there is any degree of uncertainty about their safety, or any question as to whether an adverse series of events might finally climax in harm to our society, a plan is a necessity.

Nuclear facilities are required by the NRC to have such programs. This is proof the mechanisms designed in those facilities are fallible. It is hard, from this position, for a public utility to make statements about the safety of their installation. So the plan is relegated to the category of unmentionable items.

It is not in the best interest of the pro-nuclear forces to publicize these contingency programs—or to keep the anti-forces' pot boiling by constant reference to evacuation routes.

The individuals who will be affected, therefore, need to look

223

at these plans—to examine them and consider how they might really work, as opposed to how they are supposed to work. No great expertise is required. An individual who drives on a freeway during the rush hour becomes acquainted with the problems of traffic and periodic standstills. From this experience comes sufficiently practical knowledge to review and determine the feasibility of at least the evacuation phase of these predetermined action programs.

A key part in almost every instance of disaster programming is the use of city, county, and state emergency response facilities. Local fire and police departments would be called in to assist. State agencies would be notified and placed on standby.

Where the incident demands action inside the confines of the installation, these civil units are generally required to obey the commands of the nuclear facility's managers, no matter how inexperienced these individuals might be in combatting or facing a dangerous situation. As the Browns Ferry crisis indicated, the local county fire chief was better informed and more capable of determining how to extinguish the blaze than the engineers on the operator's staff.

Outside the installation, the civil authorities are on their own, with their normal channels of command. They are trained in many of the activities required to, say, evacuate a neighborhood. But this training, in most cases, falls far short of moving a city of one million people from where they are at 7 o'clock on a given evening to somewhere else in a matter of hours. This is an almost impossible task and no amount of advance preparation will produce enough roadway and mobility to get the job done.

Few local police or fire departments have any widespread programs familiarizing their personnel with the difficulties peculiar to a nuclear contamination threat. Nor do they have the specialized equipment needed to detect various forms of radiation and protect themselves from contamination, or the medical knowledge to treat victims who have been exposed.

Present emergency plans have two specific difficulties, then. One relates to the necessity of evacuating a sizable number of people in a reasonable length of time; the other, to the use of manpower and equipment now in existence at the city, county, and state levels.

Each person living near a nuclear facility has direct access to the NRC-approved emergency plan. No special skill, other than common sense, is needed to read and make comments on the program. Here is how you can get to see the plan which potentially affects your life, and what you should do after you've read it, discussed it, and come to your conclusions as to its feasibility.

First, NRC regulations require the plan to be available in the Nuclear Regulatory Commission Public Document Room at the regional office closest to the installation in question, or in a nearby public library. A single telephone call asking about the *Safety Analysis Report* (SAR), the official title of the document containing the emergency/evacuation program, will tell you where it is.

If the facility is under construction, a preliminary SAR will be available. If the reactor is in operation, there will be a final SAR. The preliminary plan is only of marginal interest because it is usually pretty sketchy. The final version is the one which will show many items of interest.

Before reading the plan, and perhaps during your contact with the NRC, request two free booklets: One, numbered NUREG-75/111, is called *Guide and Checklist for Development and Evaluation of State and Local Government Radiological Emergency Response Plans in Support of Fixed Nuclear Facilities.* The second, officially known as Regulatory Guide 1.101, Revision 1, March 1977, is entitled *Emergency Planning for Nuclear Power Plants.* Both are available by writing to the Nuclear Regulatory Commission, 1717 H Street, NW, Washington, D.C. 20555, Attention: Office of State Programs.

The above material, combined with your knowledge of your community, will give you all the specifics you need to determine how effective the disaster programs affecting you and your family might be.

After reading the documents, three or four more telephone calls and a few letters will complete the process.

Phone the office of the chief of police in your town and request to be connected with the officer in charge of implementation of the plan, as outlined in the document. When you get him, ask a simple question: "Have you read the emergency plan for the (insert name of reactor)?"

Take the time for a few moments of conversation to see if, in your opinion, he is knowledgeable on the subject. Then call the fire department and the mayor's office. Ask the same questions.

From the information compiled in this mini-survey, sit down and write a letter to your state and federal representatives. Tell them what you found. If all personnel were aware of their responsibilities and roles, say so. But—and this is a very real possibility—if there seemed to be general unfamiliarity with the program, be direct and factual about what you discovered.

As a close to your letter, you should ask if your state's plans have been reviewed and accepted by the NRC's "Review and Concurrence" program. As of the start of 1978, only four states (Washington, South Carolina, Connecticut, and New Jersey) have programs which meet the new standards. Any regional NRC office can tell you the reasons why other states plans have failed to be in compliance. Money is one of the chief of these: no specific federal appropriations have been made available.

A study done by the Batelle Pacific Northwest Laboratory, commissioned by the Department of Energy and completed in the fall of 1977, also suggests this funding problem. In any case, some answer to the prevalent lack of funds needs to be developed. The New Jersey legislature is considering a series of taxes on utilities to cover necessary emergency costs.

Writing letters and making telephone calls is a form of political activism. There's nothing wrong with that, but few people are going to care enough to follow the outlined steps.

So the question arises about alternate courses of action. Aside from coming forward and speaking out, there isn't much an individual can do.

A major nuclear plant malfunction in one area could result in a loss of power or a lessening of available electricity in other locales hundreds of miles away. After the Browns Ferry fire, the TVA had to spend over $100 million to deliver power to its subscribers. This "imported" energy came from surrounding grids which had excess capacity. Other failures might not always work out so well. Loss of a single plant in a key area could result in widespread brownouts, and possibly force planned, pre-announced blackouts at certain selected times.

It doesn't require a lot of imagination to see everyone will suffer to some degree when a major event occurs. Those closest to the operating site, naturally, will have the worst time.

In 1977, tests were conducted on the disaster plans of two operational facilities. In January, a trial run, announced in advance, was made on the Fort St. Vrain nuclear reactor in Colorado. Then, in November, the Oyster Creek installation on Toms River in New Jersey had a simulated emergency. Neither procedure ran too smoothly or indicated a very high level of efficiency.

Accepting the fact there will be hang-ups in the disaster programs which will render them less than fully effective, where does that leave a citizen? In a difficult situation at best.

If you live near a facility, or even if you reside within a hundred miles of a plant in the path of prevailing winds, there is a remote possibility you will, one day, be called upon to evacuate your home.

That may sound like too immense a task even to think of, but as the ammonia truck incident shows, people who live in subdivisions near freeways have been evacuated. The odds are strongly against it, but it could happen.

What do you do if it does? Several experts were consulted and all agreed the key word is cooperation—cooperation with the authorities, the facility's health physics staff, and with your neighbors.

A reasonably weatherproof house is a pretty good barrier to radiation in that the radioactive elements will settle on the outer surfaces of the building instead of on you. So staying inside some form of shelter is a good first step. An automobile, with the vent system turned off, is a stuffy, but satisfactory, cover.

Most home air-conditioning or heating return-air filter systems will also remove some of the largest particulants from the air, thus reducing the radioactive threat by at least a little.

Radio and television stations, for as long as they can stay on the air, will be able to give direct advice and information to the people in an affected zone. If there is to be an evacuation, the police will do a house-to-house contact sequence in the most highly affected areas. For this, the officers will need to wear special outer garments and have access to respirators in order to protect themselves.

If evacuation becomes necessary, you should take sufficient clothing to allow for at least three changes. If the clothes you are wearing are contaminated during the time you are in the area, you will need even more. After showers and other procedures ad-

227

ministered at some central screening point, including a thorough check by various detection instruments, you will be pretty much on your own. Some form of shelter or camp will be available to those who have nowhere else to go, but the average person will be provided only with medical and health physics attention before being required to move along and free the facilities for newer arrivals.

Expect confusion because it will be everywhere. Pay no heed to rumors. Official news broadcasts from the working press will tell things pretty much the way they are. Don't expect special treatment. If you are really ill, pregnant, or have some distinct reason to believe you have been subjected to an inordinate dosage of radiation, tell someone in charge. But be ready to do the military "hurry-up-then-wait" routine, because the chaos of the moment will prohibit anyone from giving you immediate attention.

The simple rule all specialists reiterated was: Don't panic. No matter what your feelings may be, the odds are against your being in a situation of immediate, instant danger. There will be more than enough time to do anything within reason which needs to be done.

When the evacuation notice comes, pack carefully. The skin can be protected with long-sleeved shirts and long pants. Anticipate the need for warmer clothes, no matter what the season, because there probably will be some exposure to out-of-doors in the evenings and early mornings. Assuming a total loss of electricity, a battery-operated transistor radio would be a good item to take along.

Experts have had some discussion about the use of a home-made breathing filter, such as a handkerchief folded diagonally over a coffee filter and then tied bandit-style over the mouth and nose. Some feel this might help a little. Others disagree saying that it would be a waste of time as only the very largest particles would be stopped and even then, would remain in close proximity to your face and might be inhaled later.

The evacuation routes selected by the various emergency programs will most likely be jammed with cars. No one will be moving anywhere very fast. This means it will take a lot of gasoline to get you where you are eventually going. And there will be very little on sale in an area being evacuated because of a nuclear contamination threat. Cars will get onto the freeways and then, at an inopportune time, run out. This, along with the number of automobiles and the minor accidents, will cause monumental traffic problems. Some form of metering the evacuating vehicles onto the freeways will help, but it will be of little real assistance.

If all this sounds a little glum, remember evacuations to prevent exposure of the population to various industrial hazards are not at all uncommon in some areas of the United States. Many housing developments are built over or along underground natural gas pipelines, and others are adjacent to chemical plants or railway loading yards where spills can occur. Even the smaller towns are not immune. At the time this is being written, headlines in the newspapers concern a four-car derailment and subsequent problems with liquid propane gas in a quiet little city in the Midwest. Everyone in town was moved out.

As a society, we are overly inured to this. The radioactivity of a nuclear accident, though, raises other problems and could possibly have more prolonged effects, as it might take some time for an area to "cool off" enough for renewed human habitation. This is, of course, worse than a fire, which allows for rebuilding in a relatively short period of time after the event.

Estimates vary, but it appears to be generally accepted that a five-to-ten-mile radius from the center of the source will receive the most contamination. Adverse weather conditions, such as a temperature inversion or a strong wind, could cause problems for perhaps a hundred or more miles, but these would be solvable unless the release was of a magnitude on a par with the Soviet or Windscale disasters. Cleanup crews would eventually be sent in to mop up a majority of the radiation, but even then, above-normal levels would remain.

It is probable there would be places in which no habitation would be allowed for a very long time. Naturally, this would result in—among other things—a real financial drain. Imagine having your house and all of your belongings declared injurious to your health! It would be very much like starting all over again after a fire. With one big exception.

Most homeowners carry a sufficient amount of insurance to indemnify themselves against the greatest portion of a loss from fire or other natural calamity. The federal government, recognizing the problems a private insurance company encounters in protecting its clients against some acts of God such as flooding and wind damage, offers area-wide protection, through the private companies, to people who find themselves in flood zones. The Gulf Coast, for instance, has federal protection against flooding from seawater driven by hurricanes or the massive amounts of rainwater which fall during a tropical storm.

But there is no such equilateral support for nuclear accidents. In fact, the opposite is true. The federal government has fixed the limits of liability for all atomic power stations to what appears to be a very low sum.

The Price-Anderson Act, passed through both Houses in 1957, paved the way for the construction of the first reactors. Before passage of this bill, all plans to operate atomic-fueled piles were speculative. No insurance company would take the risk, and no utility would move forward without specific protection. The bill was successful enough in practice to be extended in 1975, with amendments, until 1987.

In March 1978, a federal district judge in North Carolina declared the Price-Anderson Act an unconstitutional deprivation of property without due process of law. The ruling affected the nuclear power plants only in that one district, but if upheld, it would likely affect the rest of the country. Subsequently, the U.S. Supreme Court upheld the validity of the Act, and it is currently the law of our land.

As it stands now, Price-Anderson allows total compensation to all injured parties and all loss claimants up to a specified limit. The current maximum loss is $560 million. Past this amount, both the nuclear industry and the federal government, by law, are freed of further liability.

The $560 million will, under present conditions, come from a private insurance pool of $125 million and a federal indemnity commitment of $435 million. Under today's rules, the government's portion of the payoff kitty will gradually be replaced by increased participation on the part of the private companies.

Another interesting facet of this act is the unusual method by which the utilities pay their premiums. Unlike normal businesses, which are billed in advance for the coverage supplied, the nuclear power stations are on a pay-as-the-emergency-occurs basis. This is called, grandly enough, a deferred retrospective premium (DRP), which means no one has to ante up until the first accident happens. At that point, each of the nuclear power plants will be called upon to pay into a common pool a sum, as of early 1978, yet to be set. There is an agreed-upon range, from $2 million to $5 million, but no one has fixed the specific amount.

If the amount agreed upon were to be $2 million (it might be more), then each operating reactor would contribute this number of dollars to the common fund if there were an accident resulting in damage to private properties. For example, if there were 100 on-line power stations using uranium fuel, the contribution would be a total of $200 million. The government's share would drop from $435 million to $235 million. With 218 reactors on line, the federal portion would be zero. As more units come on stream producing power, the amount remains fixed, but the available kitty moves up. Five hundred reactors would provide, at the $2 million figure, plus the insurance companies' $125 million, a total of $1,125 million coverage.

No one denies the probability of an eventual nuclear mishap which will do more than $1,125 million in damage. Sooner or later it will occur.

The current $560 million limit is tragic. It is an arbitrary number, stemming from comments made by the Joint Committee on Atomic Energy in 1965, which alluded to the Texas City disaster. (In the late 1940s, a ship loaded with nitrates exploded in the oil and petrochemical refining town of Texas City, Texas. The resulting blast and fire wiped out the city and left thousands homeless.) According to the committee, the "limitation of liability serves primarily as a device for facilitating further congressional review of such a situation, rather than an ultimate bar to future relief of the public."

That is an astounding statement. It says, in other words, if the Price-Anderson limits were surpassed by a single accident, Congress would be expected to improvise. And Congress usually does in times of disaster. But the individual would have no legal redress for more than his or her fair share of the allotted monies from the liability pool. Past this, each of us would be at the mercy of the Congress and would have to take what it might give us.

The supporters of the Price-Anderson Act base their belief in the program on history. They refer to the billion or so dollars annually being paid to miners who have contracted the dreaded black lung disease, or to Congressional action after natural disasters such as dam failures, hurricanes, or earthquakes. To date, more than a billion dollars has been voted in support to the people

231

who suffer from such catastrophes. They also recall nondisaster payments, such as the off-setting of price increases on imported oil by federal funds. All told, they have strong precedent to show our government is humane and our society sufficiently well off financially to stand such a drain.

Admittedly, Congress has been responsive to the plight of our countrymen who have been innocent victims of natural disaster. but the fixing of a maximum liability has no precedent in American business. The term "maximum" is exactly what it says. The current $560 million is the collective amount for which all parties, including the government, the various suppliers, and everyone else connected with the nuclear industry, is liable. Past this point there is no possibility of forcing further payment. So if mismanagement or malfunction causes the problem, there is no court of last resort to which an injured citizen might appeal.

Several thoughts arise from this. First, does the predetermined limitation create a careless attitude on the part of the utilities and their suppliers? Probably not, as the amount is sufficiently large to gain and maintain any company's attention. This, along with other direct business pressures, such as competition, seems to meet this problem. Second, does a limit to liability indicate the insurance companies feel there is a risk beyond computation? Again, most likely not, as they are on the line for $125 million, and certainly the actuarial tables have set premiums to cover any reasonably anticipated losses. This implies since the premiums are being paid, they are within reason and so is the probability of a mishap.

It is undoubtedly true that without the Price-Anderson Act there would be fewer suppliers to the nuclear industry. A component manufacturer would think twice before selling products for such a high-risk use. The act does protect these people. But at what cost? By any consideration it is a law which limits the ability of a citizen to collect rightful damages from those who participated in causing the harm. And as such, it is dangerous to each person in this country.

Supporters point out the record of the act to date. As of early 1976, the utilities have paid the federal government at the rate of about $90,000 per year per reactor, or a total of more than $8 million (against a $435 million loss guarantee). No claims have been paid from these proceeds. The private insurance companies have collected, in this same period, about $70 million and have paid out around $580,000. (Most of this was in payment for injuries to workers.) After the premiums have been held for ten years, something near 70 percent is refunded to the participating utilities. Refunds have been made, which speaks well for the safety of the industry.

But the Price-Anderson Act still limits the right of each of us to be indemnified from loss resulting from a nuclear event or accident. And no matter how effective the bill has been so far, it is of potential harm to each of us.

Certain major improvements are possible. If, as has been announced, the goal of the nuclear power generating industry is to become largely self-sufficient in the area of indemnity against public loss, then why not fold the insurance rebates into the program? The more than $49 million returned to the various participating companies could have gone into a fund held by the federal government. This would have resulted in a total of $57 million (the $49 million rebate and the $8 million deposited with the government by the utilities during the same time period) plus the $125 million from the insurance firms' share of the liability would have built a $182 million program.

Even better still, since the federal government is carrying most of the load, and the insurance industry was agreeable to limits on liability, why not bypass the insurance companies totally, have the utilities pay all of their premiums into a central fund, increase the premiums to $180,000 per year per reactor, and place the full burden of liability on the federal sector? Naturally, companies who build and supply components to the industry would also contribute. Had this variation been adopted ten years ago, there would now be a national fund, including interest and contributions from all sources, of more than $400 million.

Under this revised program, the federal government would continue the retrospective premium, but fix it at a set annual amount. Since an accident of catastrophic proportions is statistically unlikely, the chance of two occurring in a single year, according to an extension of the *WASH 1400* data, is highly unlikely. The chance for three in a twelve-month period is so slight as to be almost not plausible. A flat $3 million per year per reactor, up to two accidents, would add only a $6 million liability to every working unit—and the probability of having to pay the

233

full $6 million in any one year would be very remote. This reverses the previous reasoning, accepts the statistically low chance of an accident happening, and places the weight of the solution back on the people most responsible for its cause. With one hundred reactors in operation, the major accident fund would go to the base amount collected in premiums plus $300 million for the first incident, and what was left plus another $300 million for the second. This may still not be enough, but according to all calculations it should come close.

One further step is necessary because the rights of the individual citizen are still abridged by the limitation to collect on a personal loss.

Here, the federal and state governments must act together to distribute the risk proportionately throughout our society. The individual in a major city far from the reactor site, and therefore relatively safe, enjoys the use of the electricity generated as well as the prosperity stemming from the improved international position of the dollar brought about by a cleaner balance of trade. The property owner next door to the atomic power station is in much greater jeopardy.

Since, in theory at least, we all pay taxes, a very small annual amount—say two dollars per return or per family—would produce large dollar amounts for the purchase and distribution, on a state level, of the necessary emergency equipment required to meet the demands of a nuclear incident. After the initial procurement the continued funding would provide for maintenance and upkeep as well as the hiring, training, and deployment of nuclear-disaster specialist teams. This action would go a long way toward limiting the damage from a core melt or other failure which would allow the escape of radioactivity into our environment.

Alternately, each state government could purchase the necessary equipment, place selected employees into federally run schools where each attendee would pay tuition, and collect the funds from the nuclear utilities, which would then pass the charges on to the ultimate consumer.

The final phase of the revision would require the federal government and the state governments to share, on a predetermined percentage basis, in guaranteeing 100 percent of all individual losses up to amounts individuals deem prudent for their

234

needs. By working with the insurance companies, as has been done in establishing a federally funded flood program for certain areas of the United States, all persons could indemnify themselves against major loss from a nuclear accident. Insurance companies would establish low-cost annual premiums based on their statistical estimates of the damage potential. They would cover a certain degree of the total claim from the insureds, and the federal and state governments would guarantee the balance. Using the already available network of insurance agents operating in the U.S. would save us from having to set up additional employees at the federal and state levels, and would keep estimating costs and the problems of payment, customer relations, and counseling in the hands of the professionals who are doing it now.

The premium for everyone who desired to be covered would be the same regardless of proximity to an operating station. Again, this spreads the risk more equally throughout society, while assuring anyone who is concerned about recovering a possible loss a clear means of redress and satisfaction.

The key to the whole situation, of course, is to minimize the chance of an accident in the first place. We need to continue to improve the record of the nuclear industry. Providing no major technological breakthrough takes place in the next few years, the next best protection is to make certain every operating atomic station has a well-rehearsed, well-devised disaster program. This is a task for which we can all assume some responsibility.

16

■ There has been enormous progress in the nuclear field since the dark days of the 1940s. We have come a long way, and during that time very few people have been injured. The amount of radioactivity in our environment has increased, albeit slightly, but this has largely come from the explosion of military devices, not from nuclear power stations.

Our technology and understanding of the equipment needed to generate electricity safely has improved annually. But there is a lot we still do not know. A lot we still need to learn.

After everyone with a favorite alternative source of energy has spoken out, our problem will remain. Because of a single glaring fact. Our utilities are geared to produce electricity from heat.

Realistically, there is no way to rebuild our energy generation and transmission facilities to handle solar or wind or geothermal or tidal power quickly enough to save us from ourselves. Because there is a time limit. If we continue to import hydrocarbons at the present rate we are going to have to face some hard decisions—and make big changes in our basic life-styles.

Atomic energy can pull us out of the hole and keep us out long enough to work on the fusion process, which is far cleaner than the fission technique of energy production used today.

But nuclear energy is not going to save us unless we act now. Unless we do something. The price of every facility under construction has increased, year by year, so plants designed not too long ago to cost a few hundred million and come on stream in 1980 cost close to a billion and are delayed for years. Existing nuclear piles developed by companies to operate economically based specific prices for uranium find they are unable to purchase the fuel unless they are willing to pay three or four times as much. In one case, the costs went up by over 500 percent! And to date, we have not begun to address the financial problems involved in closing down a reactor site after its forty to forty-five-year operational life comes to an end. (Estimates indicate it might take four or five times the original construction price to decommission an average size station and pile.)

There are also too many unanswered questions about equipment safety and operational durability—not to mention waste disposal, transportation, and power generating technology.

We require a way out. A way to get the nuclear power stations to produce the energy we need: energy to make new jobs and assure continued employment. But we have to get it with minimum risk to our society—and our world.

In the 1950s, the Soviet launching of Sputnik I astounded the U.S. and damaged our egos. Rallying to President Kennedy's clarion call, the challenge was met. Within a decade an American walked on the moon.

Federal funds, allotted by contract to private companies for research and development of the necessary hardware, put us there, in an outstanding example of how the U.S. Government can work with industry and science to attain a predetermined goal.

A safe reactor, or at least a safer reactor with even lower emissions, is a goal we must accept as a challenge. Along with the other problems of a nuclear society, this should become the impetus for a national achievement program. It would cost far less in the end than sending men into space, and like the space program, it could produce an amazing amount of "fall-out" technology which could be used in a vast number of fields. It is also a giant step toward solving one of the most serious problems in the world today.

Somehow, it would be fitting for the largest waster of energy in the history of man to be the developer of a clean, safe, and efficient source of atomic power.

It is possible. The attainment of this goal is a bigger challenge for the 1980s than space flight was in the late 1950s—but it is within our grasp.

Along with this program ought to come the necessary revisions to the Price-Anderson Act to assure each citizen of proper redress if the ax does fall. Accompanying that should be, as discussed earlier, some form of small annual tax, either through the federal or state government, possibly collected by the utility for every present and new hookup or installation, to build a fund which would allow the individual states to purchase, train, and maintain adequate nuclear emergency teams.

If we start now, we have time. How much, no one really knows. There is going to be a nuclear accident of major proportions. It will not be anybody's fault and it will not stem from any one cause. But sooner or later, it will happen. And no responsible, informed person disagrees with this simple fact: when it hits, we have got to be ready.

Our immediate action is vital. And the monies expended will be returned manyfold to millions of Americans who will be able to continue their life-styles and provide for their families.

APPENDIX 1
WASH-1400
SOME POSSIBLE FAULT WITH THE
FAULT TREE

■ In 1878, *The New York Times* carried a brief article about electric light, citing experts who maintained that illumination from incandescence of metals could cause physiological injury to the retina of the eye because it is "habituated to a certain given intensity of illumination which is many degrees below electricity, and any material advance in intensity must, necessarily, prove irritating." This exemplifies the human habit of trying to rationalize personal stances into presentable, logical, agreement-demanding postures for the purpose of attaining support. And it shows how scientific opinion, given in good faith, can be used to create an emotion-catching argument. If you are against electric light, prove it is harmful to the human physiology and you've gained a rallying point for your position.

This same kind of scientific shuffle is being practiced by both sides of today's nuclear disagreement. Overly zealous individuals use the facts which suit them and best prove their case. And often, because nuclear physics and other branches of atomic science are complex and difficult to understand, the facts emerge as garbled half-truths and fictions.

Neither group rules out the possibility of an accident. The more rational individuals on both sides have, since the earliest stages, accepted the certainty of such an event. They have made plans and developed equipment with this very contingency in mind.

Nuclear accidents can, have, and will happen. Thus far, on most occasions, the safeguards have functioned as planned. But these devices are far from infallible, because the men who designed them are not perfect.

Recognizing this, the group favoring nuclear power takes shelter in painstakingly crafted probability sequences which

241

utilize the most up-to-date mathematical concepts to prove how unlikely a real catastrophe is.

A sterling example of this method can be found in the pages of a document called *Reactor Safety Study; An Assessment of Accident Risks in U.S. Commercial Nuclear Power Plants.* Known as *WASH-1400,* the report was compiled by a study of about seventy experts directed by Dr. Norman C. Rasmussen, who is the author of more than fifty technical papers and head of the Department of Nuclear Engineering at Massachusetts Institute of Technology. Its 3,300 pages cost about $3 million.

Aimed at reaching some conclusions about the risk of nuclear accidents stemming from problems with reactors, the report did not address itself to hazards caused by shipping radioactive materials, nuclear waste disposal, deliberate sabotage, military errors, or theft of atomic fuels. (All these are interesting possibilities and, as has been seen from the stories detailed earlier, all too real events in modern life.)

The main conclusion of *WASH-1400* is contained in the following table from the report.

Average Probability of Major Man-Caused and Natural Events

Type of Event	Probability of 100 or More Fatalities	Probability of 1,000 or More Fatalities
MAN-CAUSED		
Airplane crash	1 in 2 years	1 in 2,000 years
Fire	1 in 7 years	1 in 200 years
Explosion	1 in 16 years	1 in 120 years
Toxic Gas	1 in 100 years	1 in 1,000 years
NATURAL		
Tornado	1 in 5 years	very small
Hurricane	1 in 5 years	1 in 25 years
Earthquake	1 in 20 years	1 in 50 years
Meteorite Impact	1 in 100,000 years	1 in 1,000,000 years
REACTORS		
100 plants	1 in 100,000 years	1 in 1,000,000 years

What is being said is this: If there were 100 operating nuclear plants in the U.S., the chances are one would malfunction in such

242

a fashion as to result in the deaths of 1,000 or more people only once every million years or so. And probably only one accident large enough to kill about 100 people would happen every 100,000 years.

Reassuring numbers, indeed. There seems to be about as much likelihood 1,000 people would die from a nuclear plant disaster as there would be of 1,000 people being killed by falling meteorites.

The technique for the development of these figures is fully up-to-date. The statistical model, called a "Fault Tree Analysis," is as detailed and crafted as is possible at the present time. And the underlying logic of the assumptive chain of events is beyond reproach. The paper is a powerful representation of the technical care taken in the consideration of the modern nuclear power generating industry. Even the limitations of methodology are discussed, and it is admitted that the validity of the study is dependent upon the researcher's ability to imagine faults and problems which might occur in an operating station.

The *WASH-1400* document produced a quick reaction from the scientific community.

Space scientists and engineers, long accustomed to the differences between theoretically predicted reliabilities and the bitter fact of actual usage, pointed out no man or group is capable of thinking of every single contingency which will result in the failure of any complex mechanism. (A specially designed star tracker used in the later series of Apollo flights had a predicted reliability of 0.9999. This means there was one chance in about 10,000 the unit would fail while in use. It did. Remember Murphy's Law.)

The turmoil and discussion generated by the *WASH-1400* statistical safety evaluation will take years to subside. It almost appears both sides are having so much fun making their points, neither is anxious to quit.

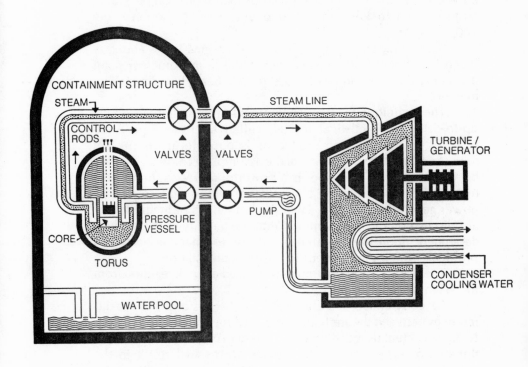

CONTAINMENT STRUCTURE

STEAM

STEAM LINE

CONTROL RODS

VALVES

VALVES

TURBINE / GENERATOR

PRESSURE VESSEL

PUMP

CORE

TORUS

CONDENSER COOLING WATER

WATER POOL

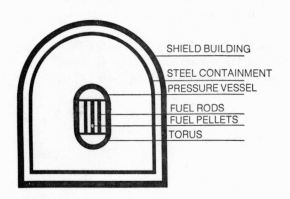

SHIELD BUILDING

STEEL CONTAINMENT

PRESSURE VESSEL

FUEL RODS

FUEL PELLETS

TORUS

APPENDIX 2
THE NUCLEAR REACTOR
HOW IT WORKS AND MIGHT GO WRONG

■ It is difficult to judge the exact state of nuclear power science.

There is much high-level scientific disagreement and lack of understanding about many of the components in today's atomic generating station. As mundane a thing as the metal used to coat the fuel rods in a light water reactor is still under investigation. Zircaloy, the material in present use, has had its operating temperatures scaled down from 2,300 to 2,200 degrees* due to recently detected stress fatigue cracking. So we are learning as we go.

A nuclear reactor is a simple enough device, as far as the physical mechanics of its operation are concerned. It is a controlled chain reaction in which nuclear energy is converted to thermal or heat energy.

The nucleus of an atom consists, in a broad sense, of elementary particles called neutrons and protons. The protons have a positive electric charge, while the neutrons have none. Very strong forces of attraction act between these particles to hold them together. But since opposites attract and like electric charges repel, the protons are constantly being forced apart. This is very pronounced in heavy nuclei, where the pushing apart of the protons weakens the nucleus. It is especially true for Uranium 235 (U 235). By bombarding the nucleus of this element with inert, free neutrons, the delicate balance is upset and it may vibrate itself apart, producing barium, krypton, and two fresh neutrons, which can and will bombard more U 235 nuclei and sustain the reaction. The fresh neutrons are moderated or slowed down in speed, usually by the use of a graphite or water barrier through which they must pass. This makes them more likely to strike more U 235 centers.

The other elements produced, called "fission products,"

*All temperatures refer to the Fahrenheit scale.

245

travel until they contact some form of matter elsewhere in the unit. When they hit, their kinetic energy of motion is changed to heat.

In the reactor, tubes filled with uranium are placed in a large vessel of water. This is called the "core." The rate of the atomic reaction is controlled by a series of rods made of a neutron-absorbing material. Called, naturally enough, "control rods," these neutron absorbers regulate the rate of the reaction. The further they are slid into the core, the more the uranium is shielded from the neutron bombardment, and the reaction slows down. The temperature of the water in the core can thus be carefully controlled. The hotter the water desired, the further the control rods are removed from the body of the pile.

The hot water is circulated through a closed loop system of pipe into a heat exchanger. This is a large vessel with coils of piping at right angles to the flow of the water coming from the core. The heat from the water in the core is passed into the water being pumped in the secondary system, and the hot water in the secondary system, in the form of steam, drives a turbine, which in turn rotates a generator.

The nuclear reaction can be halted should the operator decide to do so. The control rods of neutron-absorbing materials can be fully inserted, absorbing almost all the free neutrons thus leaving none to bombard the uranium. Or the uranium itself can be dumped or removed from the core through an electrically controlled, mechanical emergency system. Either way, things can be stopped. Up to a point. And that is the problem.

It is possible, in theory and in actual fact, for the reaction to "run away" with itself. This can happen so fast it cannot be held back. The rods can't be inserted or the core dumped rapidly enough to stop the reaction before it becomes critical.

The result is a really big mess, but not an explosion. Both supporters and detractors agree a nuclear power plant pile will not blow up. What will happen, however, is called a "meltdown" or a "core melt."

Even after the control rods are in place, heat still remains in the pile. If the core is not cooled by enough water, temperatures will rise to the melting point of the uranium, about 5,000 degrees, which is hot enough to burn through almost anything, including all the protective containment the engineers have designed around the core of the pile. The molten puddle is capable of escaping into our environment, where it would unleash a terrible amount of radioactivity into the atmosphere—and by everybody's estimates do millions of dollars' worth of damage.

All the controls—the piping system, the neutron-absorptive control-rod materials, the electric motors and pumps—are still in the process of being evaluated. Opinions on the durability of these components are constantly being revised, mostly downwards in the interest of safety.

This continuing revision is as it should be. But it is also indicative of an important fact: nuclear power generation is still somewhere in the limbo land which lies between art and exact science. And what was a "sure margin for error" today in component composition or wearability will be altered tomorrow.

A concerted, well-conducted, government-supported program focused on understanding the problems inherent in atomic power as a heat source for generating electricity is a national need. It requires a top priority in our future plans and expenditures.

No empirical work exists to determine exactly what would happen if an accident were to occur and result in a final meltdown or core melt. Estimates are based on statistical extrapolation only, and have been open to broad interpretation.

Mention was made earlier of the famous—or infamous, according to the antinuclear protagonists—*WASH-1400* (Rasmussen) report, the most quoted source for predicting the extent of damage to human health from a reactor mishap. In one sense, the job of developing the report was simple. A quick look inside a reactor will show why.

Essentially all radioactivity produced by the process is sealed inside the main reactor chamber in the uranium dioxide ceramic fuel pellets. These are, in turn, contained in special cartridges. The only way for the radioactivity to be released is by melting these pellets. So the only accident which would produce huge emissions would be one where the pellets were "melted down"—hence the word "meltdown," used to describe the ultimate class of reactor accident. Since normal operating temperatures in the unit are in the 1,000-degree range, and the melting point of the uranium ceramic is about 5,000 degrees, core melts will not be caused by minor problems in the system.

There are, in fact, only two ways such an accident could be

produced. First, the neutron-dampening control rods might be withdrawn too far or too fast. This is the least serious possibility, as several self-regulating factors limit an occurrence of this nature. First, the hotter the uranium gets, the harder it becomes to heat it further. It begins to absorb the heat-producing neutrons, thus lowering the atomic bombardment. Second, the water coolant bath will, when the capsules get hot enough, turn into steam, eliminating one of the better neutron conductors. Finally, the control rods themselves have an emergency operating system, based on stored hydraulic pressure or, more simply still, gravity, and they will automatically slam back into the pile at the first signs of a heat buildup. This emergency insertion process is referred to in the trade as a "scram" and the need for it in the system is more than for emergency situations. Electrical equipment is very reliable. But about once a year, as an average, some component of the generator goes bad, resulting in the removal of the unit from the power-producing line. When this happens, the steam turbine has to be shut down, causing a strong amount of back pressure in the lines and ending up by producing bubbles of steam in the area of the pile. These in turn cause an increase in the power from the reactor, and the emergency system is called upon to scram or insert the control rods. If the scram system did not operate reliably, a type of accident, called ATWS—for anticipated transient without scram—would occur, and there would be danger of a meltdown.

In reality the scram systems do work and are used in the normal operation of the unit. But they are mechanical, in the sense that the rods must actually be moved in and out of the pile, and there is some possibility, no matter how remote, a system might fail. The determination of the incidence of this happening was one of the two main areas the WASH-1400 study had to focus on.

The second, and according to most, more dangerous difficulty is the loss-of-coolant accident (LOCA or LOC), in which the water is suddenly boiled away into steam, because of a leak in the system.

The water used in reactor cooling is heated by the process to around 600 degrees, and must be kept under pressure or it would boil off. If one of the pipes, carrying internal pressures ranging from 1,000 to more than 2,000 pounds per square inch (psi), were to spring a leak, the water inside would instantly escape into the air as superhot steam. This situation is known as a "blowdown," and would leave the reactor without water. Again, the absence of the water would shut off the chain reaction, but the radioactive decay process in the uranium dioxide fuel pellets would continue and the temperature would skyrocket upward to the 5,000-degree melting point in seconds. In pressure-type water reactors, fuel without its proper water coolant would reach a point of no return in thirty to forty-five seconds. An emergency addition of water could, because of the heat and a potential reaction with the zirconium material used to clad the fuel pellets, actually cause an increase in the temperature and bring things even closer to the meltdown stage.

Complete meltdown of the fuel would not take longer than a half hour, and after around sixty minutes, according to most estimates, the now-liquid uranium would burn through the shell of the reactor vessel, releasing an astonishing amount of radioactivity into the atmosphere. No atomic explosion would take place, but a blowdown could lead to widespread nuclear contamination.

A number of tests and special full-time sensors which detect rises in humidity caused by the release of steam have been designed into every reactor. But more than a little controversy, not all from antinuclear forces, centers on this crucial possibility.

Responsible engineers accept the idea of a leak going undetected until it becomes an irreversible problem, and extend this thinking to agreement that a loss of coolant could happen. Their answer to this lies in the design of a secondary, redundant backup system, called the "emergency core cooling system," or ECCS. But even this safeguard is based on the assumption the heavy steel reactor vessel is not ruptured and the loss of coolant has resulted from a leak in a pipe.

Probability, then, is played in the design of safety backups, because no team of engineers can or will agree to perfection in any mechanical or electrical device.

The emergency core cooling system is another odds-on favorite to work. Failure of valves or pumps has been minimized by redundancy in design, and even pessimistic estimates give a 99 percent reliability factor to the system as far as its actually being able to deliver water. Where things fall down is in the question of whether or not the water, when delivered, will do any good.

There has been extensive controversy over whether the emergency coolant, delivered by the almost certain ECCS, would stop the meltdown.

By the time the replacement water fills the reactor vessel to the lowest level of fuel pins, the core will have gotten very hot. The liquid will flash into steam and the pressure will retard further flooding to a rate of about one inch per second. (Remember, in thirty seconds from the time of coolant loss, the process will have gone too far.) An explosion, caused by the sudden buildup of steam pressure, could happen. During this period, while the vessel is filling, the only cooling effect is caused by heat transfer through the steam.

Not too much is understood about this, and scientists are quite uncertain what the cooling rates of the water and steam mixture would be in such a case.

Research has been done, but it was based on the use of mock-ups instead of a real reactor. In the testing, rods the size of fuel pins were heated electrically and subjected to the water-steam cooling bath. Readings allowed the development of several theories on the effectiveness of this form of coolant, and these in turn were pushed through a computer to arrive at operational criteria for the emergency cooling systems.

Everyone agrees this method is error-prone and inaccurate. Large margins were added for increased safety and the results incorporated in the final designs. But no one is certain how large the special "ignorance margins" should be. In fact, a group of noted scientists and engineers has, since the development of the original standards, conducted its own experimentation, and determined the leeway was not sufficient. This team, headed by Henry Kendall, felt so strongly about the matter it made the data public and submitted it to the AEC.

The result was a year of hearings in which many AEC-associated scientific and engineering personnel sided with the Kendall team. The final results called for an increase in the safety factors and the curtailment of the operation of some reactors until these could be incorporated.

But the matter was far from settled. Kendall and his crew,

while gratified at the action, were still upset by the lack of positive knowledge. Their agitation has finally resulted in the scheduling, during 1977, of elaborate tests using an actual reactor specially constructed for this purpose.

With such open dissension among the design and operating staffs, it is little wonder the WASH-1400 report was received with less than instant belief and respect. But a further area of disagreement clouds the issue even more.

Assume the loss-of-coolant accident did occur and the emergency system was unable to restore the situation. A meltdown would take place. But what then? A large puddle of superheated fluid would rapidly form on the bottom of the main vessel. This is a thick-walled steel unit, and it would take a few minutes for the hot matter to melt its way through into the next containment barrier, a structure made from heavily reinforced concrete and lined with sheet steel plate. The most frequently used containments are tested to withstand internal pressures of more than five times normal, and are designed to be strong enough not to fracture under impact from wind-blown objects coming at hurricane force, runaway automobiles, or charges of various sizes of conventional explosives. It would take a really large chemical bomb to crack this massive protective shield.

The internal walls of this construction would be cooler than the floor area where the molten fuel would finally congregate, and aided by a number of built-in air ducts, the largest amount of the radioactivity released by the molten pile would plate or coat these surfaces. Special sprays, also set to work in an emergency, would wash the radioactive iodine from the air, returning it to the walls and floor. If this containment device could withstand the intense heat for even a few hours, most of the radioactivity would not be passed into the atmosphere, but, rather, would be held inside the thick structure.

Finally, though, without question, the hot puddle would succeed in melting its way free of the restraint of the container, releasing toxic radioactivity.

The situation would be much worse if the containment should fail to withstand the heat and pressure for several hours. Truly mega-amounts of radioactivity would be freed into our environment and, depending upon the weather conditions, would wreak a variety of harms.

251

In the *WASH-1400* study, probabilities were assigned for each of the main types of failures which could result in a loss of coolant, and produce a meltdown. Extrapolation of the data in the report, accepting all the base material from which the statistics were derived, is interesting. A meltdown occurrence, according to the report, is almost a certainty. The probability it will occur more than once for every 20,000 reactor years is low, but the chance of it happening one time during that period is very good indeed.

It might be well to note a "reactor year" is one year's operation of a reactor anywhere. Twenty reactors operating one year equal 20 reactor years. One thousand in operation for the same period equal 1,000 reactor years. It would take an estimated 400 nuclear reactors to produce 100 percent of the electrical energy needed to meet this nation's needs. If all 400 were in operation, we could expect the dreaded event to occur once every fifty years. Some people, in other words, would have two meltdowns and the subsequent release of radioactivity occurring in their lifetimes. And almost every person would be alive during one such occurrence.

But all this assumes the NRC is right in its belief the emergency core-cooling system will function as planned and not as the group of concerned scientists maintain it will. If we alter the *WASH-1400* figures to meet the higher limits for error imposed by this collection of experts, the probability of a core melt increases about threefold, to one meltdown occurrence for every 6,000 reactor years. With the same 400 power plants in use, this would mean there would be a core melt about once every fifteen years.

That is getting a little frequent for anyone's health.

With 200 reactors in operation, we could anticipate such an incident once every thirty years—also a little too often.

We do have some history on long-term reactor operation. Those belonging to the U.S. Navy and maintained without thought of costs have now accumulated somewhere near 2,000 reactor years—and there never has been a core melt. Civilian units, onstream much later, have insufficient operational hours to pass judgment. To date, there have been less than 300 reactor years of experience.

Since the occurrence of an incident is almost assured, at least statistically, the real question must be how much damage

252

such an event would cause: How many lives would be lost and how much property destroyed?

According to *WASH-1400,* and based on the probability of 20,000 reactor years per meltdown and 400 operating reactors, a core melt would occur each fifty years. The annual death rate (total deaths caused by the incident divided by 50) would be on the order of two-tenths of one death per year from acute radiation sickness and ten eventual cancer deaths from contamination related to the event. In addition, a property-damage and cleanup cost of about six million dollars would be anticipated. This does not include incidents brought about by deliberate sabotage or terrorist action.

The difficulty with these figures lies in the evacuation concept utilized by the *WASH-1400* study panel, and, as discussed earlier, in the more pessimistic numbers for more frequent reactor failure. Combining these, experts have estimated the annual death figure to be more on the order of 600 persons.

The death of 600 people annually is no inconsequential matter. But place it in this perspective:

Projected Annual Numbers of Deaths in 1977-78, (12 months)

Traffic*	45,000
Burns From Non-Traffic Related Accidents	8,000
Aspirin and Aspirin Compounds	200
Electric Shock	1,000
Lightning	160
Choke on food	3,000
Aircraft Accidents	2,000

*45,000 people will be killed in traffic-related mishaps. Another 2,000,000 will be injured.

This table was compiled from various federal sources and represents a fair sampling of the dangers of modern life. It does not include deaths from industrial incidents.

With this kind of annual death toll, 600 doesn't seem so high. It's not wise to add another increment to the accepted levels of unnatural death, but the alternative—from the standpoint of our energy-consuming society—is probably far worse.

Can we live with such a figure? The answer is obvious. We already do. And no one seems to notice.

Another much-discussed cause of potential injury from the atomic industry relates to the transportation of radioactive materials from place to place, and the matter of nuclear waste disposal.

The above mentioned possible sources of harm are serious, but combined, the most pessimistic projections of illness and death do not exceed one or two people per year. And figures on illness from routine emissions which might occur from power stations add about another ten lives to that total. This may appear a harsh way of seeing the matter, but taken at its worst, the nuclear industry, with all reactors generating 400 kw of electricity annually, would probably be responsible for taking less than 620 lives per year. The additional increase in danger for the average person in our society would be equivalent to riding another ten to fifteen miles per year in an automobile.

All this sounds logical. But when the accident occurs, and all the 600-per-year deaths come in one swift cycle, the strain on human emotions will be greater than any of us has ever before experienced. It is one thing to read about thousands killed in a flood or hundreds dead from an airline crash, but quite another to have a man-made occurrence which kills and injures individuals who were otherwise safe in their homes. The backlash from such an incident will be severe.

Things can be done to make the nuclear power industry even safer than it is today. But there are difficulties in accomplishing the necessary changes and technological advances which must come about.

The situation is far from hopeless. Research is moving rapidly forward and new ideas are being tried on a daily basis. Stumbling blocks still exist, though. There is heavy pressure to nuclearize the electrical generating industry as an answer to this country's energy requirements, at a rate faster than prudent. And public attitudes, which have undergone massive changes since the 1950s, are causing re-evaluation of the worth of the risks presented to our society from this new source of power. Nuclear energy is dangerous. So are coal mining, and petroleum refining, and other major industrial processes. But none of these cause the same emotional reaction as atomic power.

Without downplaying the danger, it will be helpful if we adopt a more rational attitude toward this new industry. Like it or not, it looks as if it is with us to stay.

APPENDIX 3
THE EFFECTS OF RADIATION ON HUMANS
YOU CAN'T SEE IT, SMELL IT, TASTE IT, OR FEEL IT— BUT IT'S THERE

■ Radioactivity and exposure to radiation are accepted by our society as being extremely detrimental to human health. But there are widespread misunderstandings about the danger.

Radiation is not a new or exclusively man-created phenomenon. Radioactivity occurs in nature. It surrounds us all, every day. So from purely natural sources there is a significant daily exposure to every man, woman and child on earth.

Man's addition to worldwide radioactivity is from the regular production and use of large amounts of radioactive material and the subsequent massive dose-exposure potential inherent in this practice. This again is not new. We have done the same kind of thing with a broad spectrum of drugs and chemicals. Substances with proven long-term lethal effect are used in our daily environment.

Radiation is no better or worse than these more familiar poisons. But since it can't be heard, seen, felt, or smelled, it falls outside the ken of our normal senses and takes on a terrifying mystery—an added emotional force.

Most experts are not clear about the effects of low-level dosages of radiation on human health. But upper-limit dosages are well understood from direct observation of individuals who have fallen victim to this uniquely modern phenomenon.

Two organizations, each with ample prestige and expertise, have issued independent reports which correspond on significant points. The first, the National Academy of Sciences's National Research Council Committee on Biological Effects of Ionizing Radiation, is an American organization. The second, the United Nations Scientific Committee on Effects of Atomic Radiation, is international in its scope and membership.

Additionally, ongoing surveillance by two other groups,

mostly interested in developing standards for maximum permissible dosage levels, provides enough information to allow for reliable conclusions of the effects of radiation.

Natural radiation exposure is not slight. And like all other forms, it is cumulative. To understand how much we get on an annual basis, a standard measure was needed. Called a rem, it can be likened to a foot or an inch or a second. We can tell someone how far it is between two points in feet, or how long something is going to continue by citing seconds; The rem tells us how much radiation. One rem equals the amount of any type of ionizing radiation which will produce the same damage to humans as one roentgen of approximately 200 kilovolts of X-radiation. That's really a confusing definition. Suffice to say, a single rem is a lot of radiation, and for purposes of measuring amounts tolerable to humans, it is necessary to divide that by 1,000. This new unit is called a millirem (mrem).

Each person in the United States is exposed to about 130 millirems per year from natural causes. But natural radiation levels vary, and how much you may get depends on where you live, and how.

Natural radiation exposure is higher for those people who live on or inside stone or rock concentrations because the radioactive materials are suspended in the rock material. In Texas and Louisiana, for example, the exposure level is about 100 millirems per year, and in Wyoming and Colorado, mountainous states, the level increases to 250. Living in a brick or stone house adds about a 30-millirem-per-year exposure to an individual over what would be received by living in a wooden home. Measurements also indicate a person living in Brooklyn, which is built on sand, receives around 10 millirems per year less than a Manhattan-dwelling counterpart because the island is composed almost totally of granite.

To give some perspective to this picture, a medical or dental X ray will add between 50 and 100 millirems' exposure, on the average. So the natural radiation dosages are very low. But there are places on earth where they are significantly higher, and the people who live there provide us with a growing body of reliable information.

Specific areas in both India and Brazil, where there are monazite sands with high concentrations of thorium and uranium,

258

have populations which are annually exposed to radiation of 1,500 millirems, a level over ten times greater than the U.S. average. Continuing studies of these groups indicate no unusual effects from this radiation level, so it is a relatively safe assumption the human physiology can withstand the 1,500 millirem-per-year level without discernible harm.

Radioactivity, then, is a natural hazard, to the extent each person undergoes some level of exposure. But the concentration is in one-thousandths of a rem.

Man-made exposure levels are considerably higher.

For instance, nearly 15,000 people in the United Kingdom were given X-ray treatments for a type of spinal arthritis. The average whole-body dose for the group was about 400 rems—not millirems, but full rems. A statistical analysis of the group reveals there have been about 100 excess deaths—that is, fatalities over and above the norm for a given age, occupation, and so on.

Perhaps the largest, most carefully studied groups are the survivors of the Hiroshima and Nagasaki atomic bombings. About 24,000 people received dosages in the 130-rem range and more than 100 cancer deaths have resulted. Yet no evidence to date indicates an individual's susceptibility to other diseases is affected by the high dosages. There is no statistically higher incidence of death from illnesses aside from cancer, than attained by control groups.

Another very heavily exposed group is composed of about 4,000 uranium miners who received ultra-massive dosages in the 5,000-rem range from the inhalation of radon, a gas given off by the uranium while being mined. Cancer deaths among these workers have already exceeded 100 persons, and there are, regrettably, likely still more to come. Other less dramatic groupings include the 50 excess cancer deaths among the 775 American women employed in the task of painting radium numerals on watch dials to make them readable at night. This group was exposed between the years of 1915 and 1935, and have been followed for great portions of their life spans. And finally, over 900 Germans received radium treatments for the same type of spinal arthritis the British tried to cure with X rays, and there are several groups of ten or so people which have shown evidence of excess deaths due to cancer after high-level exposures.

One quite interesting fact obtained from all these studies

259

is the knowledge that the body has mechanisms for repairing damage caused by exposure to intense radioactivity. Chromosomes broken by radiation have been observed under a microscope to reunite themselves. And while single-shot dosages in the magnitude of 1,000 rem, given to laboratory mice, will kill them, the same amount of radiation spread out over a few weeks has no immediate effect.

One other specific thing has also become known and it serves as the basis for radiological treatment of certain types of cancer. Cells which multiply rapidly are far more susceptible to damage from radiation. Since most cancers are composed of rapidly dividing cells, the radiation treatment affects the malignant portions of the body more than the surrounding tissues.

The difference between a full rem and a millirem is a thousandfold, and between 1,000 rem and one one-thousandth of a rem the difference is a millionfold. So the radiation levels which cause cancer in the human body are much, much higher than any to which we are naturally exposed, even over a lifetime.

Genetic effects is another questionable area. Again, these occur naturally. The question is: How much does radiation increase the problem? About 3 percent of all live births exhibit some form of genetic defect. Actual work with humans in this area is very limited, and most of the standards have been set by working with animals. Survivors of the Japanese atomic bombings have been studied to determine the presence of birth defects in their offspring; this sample has produced no evidence to indicate a trend toward a greater number of such defects. Both major organizations which study these matters, mentioned earlier, give very low probabilities to this, based on relatively high exposure levels.

The radiation exposure standards set by the Nuclear Regulatory Commission call for no member of the public, including those living closest to an atomic reactor, to receive an annual whole-body dosage of more than an extra 5 millirems per year, or a maximum of an extra 15 millirems per year to the thyroid. The additional 10-millirem-per-year thyroid dose is due to the pronounced affinity this gland has for iodine, and the presence of radioactive iodine in the by-products of an atomic pile. This level is about one-twenty-sixth more than the national average and far less than the exposure difference which would occur if a person moved from Texas to Wyoming—or, since the atmosphere protects us from radiation, less than one would receive flying at 30,000 feet for about an hour.

The NRC regulations also cover the reactor coolant being released into streams. In addition to being restored to a temperature which will not prove harmful to marine life, the water must be low enough in radioactivity count to give a maximum five-millirem exposure to a person who derives 100 percent of his drinking water and fish from the release area and also swims in the discharge for a full hour every day.

Expert opinion, however, indicates there is more risk of contracting a serious radiation-induced illness for workers in the military or in private industries not associated with the reactor business. Many radioactive materials are used under poorly controlled conditions in a number of manufacturing and inspection processes.

Recent news stories, for instance, about the possibly damaging effects of long-term, low-level radiation on workers in the Portsmouth Naval Shipyard in New Hampshire, where nuclear vessels are repaired and overhauled, and to some members of an Airborne regiment involved in "Operation Smokey," a 1957 maneuver which featured an atomic bomb explosion, have brought new focus on the problem. Government studies are now underway in an effort to develop better understanding of this still-nebulous area. But it seems likely we will be so far embarked on our atomic course by the time meaningful results are available we will have no way to stop. Additional information may allow us to further safeguard the population, but it is difficult to imagine even the most pessimistic reports will cause us to abandon our nuclear efforts.